1986

The Body as Property

The Body as Property

RUSSELL SCOTT

THE VIKING PRESS   NEW YORK

Copyright © 1981 by Russell Scott
All rights reserved
First published in 1981 by The Viking Press
625 Madison Avenue, New York, N.Y. 10022
Published simultaneously in Canada by
Penguin Books Canada Limited

Library of Congress Cataloging in Publication Data
Scott, Russell.
The body as property.
Includes index.
1. Donation of organs, tissues, etc.—Law and legislation.
2. Body, Human—Law and legislation.   I. Title.
K3611.T7S36   344′.0419   80-52003
ISBN 0-670-17743-1

Printed in the United States of America
Set in CRT Times Roman

TO

my wife, Dorian, and my son, Fergus,
not least because they were perfect
companions in the English village of
Steep, where this book was written.

Acknowledgements

My Publishers have generously encouraged me to record here my gratitude to those whose enthusiasm and advice have determined the form and much of the content of this book. I am delighted to do so. My friends Alan Scholefield and his wife, Anthea, of Hampshire, England, both busy authors, were responsible for my embarkation on the project, and by unselfishly sharing their time and literary experience during two calendar years enabled me to keep on course. I have been "twice blest" by the friendship and judgement of my agents, Elaine Greene of London and Roberta Pryor of New York, and their profound knowledge of the world of books. I wish also to thank my transatlantic editors, Elisabeth Sifton and Philippa Harrison, for their skill and forbearance. In Australia I have been the beneficiary of the unflagging interest and erudition of Mr. Justice Kirby, Chairman of the Australian Law Reform Commission, and the thoughtful advice of Mr. George Brouwer, the Commission's Director of Research. Other Australians to whom I am indebted include lawyer Mr. James Spigelman, Dr. Donald Sheldon, a specialist surgeon who has directed me on many occasions, Drs. John Stewart and David Tiller, physicians extraordinary, Dr. Kevin Lafferty, whose work in immunology is widely known, obstetrician Mr. Ian Johnston, who delivered Australia's first "test-tube" baby, and Dr. Tom Connolly, a moral theologian of deep humanity. In the United States and England I was provided with courteous and unhesitating assistance by many medical men, academics, lawyers, coroners, and others, and also by institutions. In particular I should thank Englishmen Lord Smith of Marlow, President of the Royal Society of Medicine, and internationally known transplant surgeons and physicians Professor Roy Calne of Cambridge, Mr. Terence English of Papworth near Cambridge, Mr. Maurice Slapak of Portsmouth, and Dr. Roger Williams of London, together with pioneers in *in vitro* fertilization Mr. Patrick Steptoe and Dr. Robert Edwards, both of Cambridge, and lawyer Dr. Peter Skegg of New College, Oxford. American lawyers were especially helpful and included Justice John P. Flaherty of the Pennsylvania Supreme Court, John W. Murtagh, Jr., of Pittsburgh, Regina Rockefeller of Boston, and Morris E. Burton of Frankfort, Kentucky. Invaluable assistance came from the American Medical Association, American and British transplant registries, government instrumentalities, blood banks (in particular Hudson Valley Blood Services of Valhalla, N.Y., through Mr. Stephen Nestanpower), tissue banks (such as The Living Bank, Houston, Texas), and the Division of Artificial Organs at the University of Utah, whose Dr. Robert L. Stephen provided advice on prosthetics. I am pleased to record that I received prompt and detailed information to my formal inquiries of the Council of Europe (General Secretariat—Division of Legal Affairs), and the governments of Belgium, the Federal Republic of Germany, France, Greece, the Netherlands, Switzerland, the United Kingdom, and the United States. As well, I can attest to the remarkable efficiency and courtesy of the staff of many libraries, including those of Harvard University Law School, the New York Bar Association, the Institute for Advanced Legal Studies (London), the London School of Hygiene and Tropical Medicine, and the Australian Law Reform Commission, together with the Wellcome Library (London), and the public libraries of London and New York.

Contents

The Body as Property

CHAPTER 1

The New Human Body

In May 1968, a local newspaper in California carried the following advertisement:

> NEED A TRANSPLANT? Man will sell any portion of body for financial remuneration to person needing an operation. Write Box . . .

On Sunday, January 26, 1969, the *Los Angeles Times* carried the following advertisement:

> EYES for sale or transplant. $50,000 each—help someone you care for see and in return you'll be helping others. Only sincere parties apply please . . .

West German newspapers in March 1972 reported an offer by a thirty-one-year-old father of two small children to sell one of his kidneys for a million marks. His job was keeping him almost permanently overseas and he intended to use the proceeds to settle in the German countryside with his wife and family. Hospital tests had confirmed that his kidneys were healthy, and he was "confident of finding a buyer."

In December 1976, Scotland Yard detectives were investigating reports of extensive trafficking in human body parts in London. A few weeks later, in January 1977, a bank official in England publicly offered to sell one of his kidneys to any Arab in need of one for £8,000. Perhaps he was aware that English transplant surgeons had by then become accustomed to the arrival from the Middle East of patients requiring kidney transplants who bring their kidney donors with them. In 1979, a visitor-patient from Pakistan offered an English surgeon £12,000 to perform a kidney transplant. For a kidney, the patient was going to pay his donor about £400, approximately one-thirtieth of the fee offered to the surgeon. The surgeon decided not to do the operation, but others have done the opposite. Transplant teams in other cities around the world have received similar requests when visitors accompanied by living donors have arrived from countries that lack the facilities for organ transplantation.

The Sunday Times of London, in a detailed report about worldwide trafficking in human organs, disclosed in 1977 that between 1970 and 1976 a South Korean medical practitioner had sold twelve thousand pairs of fetal kidneys to an American medical supply corporation at an average price of $15 a pair. They were in great demand because fetal tissue grows more quickly than adult tissue and is therefore more efficient as a medium for developing cultures. The kidneys were taken from aborted fetuses (with written consent from the mothers), packed in ice, and flown out of Seoul the same day. A spokesman for the corporation which bought them said that it acquired fetal tissues from some 250 sources in 12 countries, and that the kidneys, which were in great demand, were resold in the United State to various laboratories and hospitals doing research aimed at producing antivirus vaccines. The seller later added that there were a number of Korean suppliers of fetal kidneys and that he was by no means the main one. No illegality was involved because abortion was lawful in South Korea and the laws of neither nation prohibited payment.

British commercial transactions in fetal material received limited official approval in 1972, in the report of an advisory group set up by the government under the chairmanship of Sir John Peel "to consider the ethical, medical, social and legal implications of using fetuses and fetal material for research." "Since 1968 commercial use of the placenta and retroplacental blood, not otherwise used by the National Health Service, has been accepted practice," the report said, "provided that the products to be derived from them are intended for therapeutic use. We see no ethical or legal objections to this practice." Financial transactions concerning fetuses should recover no more than the necessary costs involved, it went on, and "in no other circumstances should there be any monetary exchange for fetuses, fetal tissue or fetal material." More significantly, the Peel Report emphasized the great importance of fetal tissue in many areas of medical research: "in preventive medicine there is generally no practical substitute for the fetal tissues used, [and] it seems probable that the use of fetal tissue will offer the only chance for growing the viruses thought to cause hepatitis and infantile gastroenteritis."

Whether or not these events possess a deeper human significance than may appear at first glance, they are all unmistakable responses to a remarkable new social fact. Dead or alive, the human body now has an intrinsic value. To be precise, that value inheres not in the body as an entity but in its component parts.

It could be inferred that there is a recognized market in human body parts. Recent fiction and film publishers have adopted this notion. For example, one American novelist in 1977 priced a "four-tissue-match" kidney at $200,000, and a "four-tissue-match" heart at $1 million.* At least this is an imaginative improvement on the traditional scientific comment that the human body, if reduced to its basic chemicals, would have a market value of only a dollar or two.

In contrast to the United States, European nations outside the Communist bloc are virtually unanimous in treating as illegal the sale and purchase of human body parts for profit. A 1975 report of the Council of Europe carefully examined the attitudes of fifteen of its twenty-one member states. The report found that, apart from blood, trade in renewable or regenerative body materials was "almost universally" illegal, while trade in organs and other nonregenerative tissues from living bodies was illegal in most of the member nations. As for dead bodies, neither the body itself nor any of its contents could be lawfully sold in any country except Greece either by the deceased before death or by the deceased's relatives or anybody else after death. Greece plugged this loophole in its laws with a new statute in December 1978.

---

* "Four-tissue-match" refers to the optimum possible matching of the tissues of two persons under present medical knowledge. There are known to be at least four sites on the genes of every person which display specific sets of characteristics (proteins) which identify that person's "physical self." When all four sites on another person's genes are compatible or similar, this compatibility (which is thought to result in a lesser chance of rejection in transplantation) is often called a "four-tissue-match."

Other nations that have set their faces against such trade are to be found in the Western Hemisphere, Australasia, and southern Africa. In fact, speaking generally of the Western world, it seems that financial dealings in human organs and in most tissues are at best private and at worst clandestine or illegal.

Over the past three decades new uses for human tissues have made the body in a sense priceless. Its worth is self-evident without being money-related, although it would no doubt command a high money price if it were traded in an open market. Many of its components are in short supply and great demand. How can a rational estimate be made of the value of an organ that, if successfully transplanted, would save the life of a dying patient? What should be the price of an extract from a gland that will enable an infertile woman to become pregnant or will make a dwarf child grow?

Is it ethical to attempt to calculate the value of the human body in money? Many take the view, to use an old legal phrase, that human tissues should not "sound in money." Yet it would be misleading to say that the human body has never in the past commanded a money price or had some other kind of value. Even in this eerie territory there is little new under the sun.

For centuries, cadavers have had a traditional use in medical schools as part of students' education in anatomy, and in museums for display. Anatomical dissection was practiced in European universities as early as the fourteenth century. The requirement for bodies was limited, and the number needed, as a proportion of the dead, minuscule. Even so, a vigorous trade in human corpses for anatomy schools in the early nineteenth century in Britain gave rise to the famous "body-snatching" scandals that culminated in multiple murder, public uproar, and the passage of the Anatomy Act of 1832. This Act remains in force to this day in England and has been widely copied throughout the world as a code for regulating the practice of anatomical examination. Some of the basic procedures it outlines for donating dead bodies were borrowed by the first British-style transplant laws of the 1950s and 1960s, and these too still apply in a number of nations.

Obtaining bodies for anatomy was never a legal or social problem in medieval or postmedieval continental Europe, where more enlightened attitudes prevailed than in Great Britain and Ireland. The traditional European approach, seen in France, Germany, Italy, and

Austria, was to require the licensing of anatomy schools and to allow the unclaimed bodies of persons who had died in public institutions to be used for anatomical study.

The first official British measure was a charter granted by the Town Council of Edinburgh in 1504 and ratified by a law of the Scottish King James IV on July 1, 1505, whereby the Guild of Surgeons and Barbers was given the right to claim the body of one executed criminal each year for anatomical dissection ("ane condampnit man efter he be deid to mak anatomea of "). Edinburgh thus became the cradle of anatomical study in the British Isles. This method of supplying cadavers to anatomy schools was also adopted in England in 1540, when Henry VIII, in the thirty-second year of his reign, signed the first law that granted to the United Company of Barbers and Surgeons the corpses of four executed felons "yerely for anatomies." Apart from an equivalent grant by Elizabeth I in 1564 in favor of the College of Physicians, the government closed its eyes to the whole gruesome business for almost three hundred years.

During the eighteenth century, the number of medical students in Britain greatly increased, and private medical schools began to appear. These schools trained students for examination by the Company of Surgeons (which divorced itself from the Barbers in 1745), the College of Physicians, and the universities. Because of legal restrictions on the supply of corpses for anatomical dissection, a severe shortage developed. A new species of tradesman appeared, the "body snatcher" or "resurrectionist," who was willing to dig up newly buried corpses and sell them to the anatomists. The trade became big business and soon attracted criminals. Owners of private medical schools were coerced into paying retainer fees and hush money, and began to receive threats of blackmail and violence. It is recorded that one Joshua Brookes, who in 1814 conducted the Blenheim Street Theatre of Anatomy in London's West End, was suitably dealt with for refusing to pay a five-guinea fee demanded by some resurrectionists. On the same night two dead bodies in a high state of decomposition were dropped by the men he had offended in the street near his premises. Two young women stumbled over one of the bodies and "raised such a commotion that had it not been for the police Brookes would have fared very badly at the hands of the mob which soon collected."

By the beginning of the nineteenth century, great numbers of bodies were being stolen from graveyards and no corpse was safe from disturbance, no matter how eminent the deceased. Cases were reported of medical students attending the funeral of a relative and finding the body on the dissection table a day or two later. Anatomists who required a particular cadaver as an interesting specimen could pay a suitable fee and be sure of getting it. One famous surgeon, Sir Astley Cooper, said in evidence to a British Parliamentary Select Committee in 1828 that there was no newly buried person whose body he could not obtain, "let his situation in life be what it may."

The extent to which body snatchers became a fixture in British society was shown by Charles Dickens in *A Tale of Two Cities* in the person of his immortal "resurrectionist," Mr. Jerry Cruncher:

> Mr. Cruncher and his son were walking down the street—"Father," said Young Jerry as they walked along . . . "what's a Resurrection Man?"
>
> Mr. Cruncher came to a stop on the pavement before he answered, "How should I know?"
>
> "I thought you knowed everything, father," said the artless boy.
>
> "Hem! Well," returned Mr. Cruncher, going on again, and lifting off his hat to give his spikes free play, "he's a tradesman."
>
> "What's his goods, father?" asked the brisk Young Jerry.
>
> "His goods," said Mr. Cruncher, after turning it over in his mind, "is a branch of Scientific goods."
>
> "Persons' bodies, ain't it, father?" asked the lively boy.
>
> "I believe it is something of that sort," observed Mr. Cruncher.
>
> "Oh, father, I should so like to be a Resurrection Man when I am quite growed up!" . . .
>
> Mr. Cruncher added to himself: "Jerry, you honest tradesman, there's hopes wot that boy will yet be a blessing to you, and a recompense to you for his mother."

The development of this state of affairs was obviously assisted by the failure of the law to provide practical help beyond some ancient principles protecting graveyards, and a curious common law rule that no one could have property rights in a human corpse. A corpse was therefore incapable of being stolen because the law of theft operated only to protect rights of property and ownership. There could be property rights in a coffin or in a shroud, so that theft of these might result in imprisonment or even execution. However, according to one of the most august sources of English law, Blackstone's *Commentaries on the Laws of England,* published in 1765, "stealing the corpse itself, which has no owner (though a matter of great indecency), is no felony." French law, by way of contrast, "directed that a person who had dug a corpse out of the ground in order to strip it, should be banished from society, and no one suffered to relieve his wants, till the relations of the deceased consented to his readmission."

One of the most extraordinary examples of the no-property rule was a nineteenth-century case in which an eighty-four-year-old man was charged with theft after incinerating the corpse of a dead baby whose father he was said to be. He had put the corpse in a ten-gallon cask of petroleum and set fire to it. The court decided that the corpse could not be the subject of property and that the theft and burning of the body did not amount to a crime.

The law of the eighteenth century did, however, protect consecrated burial grounds and punished certain minor misdemeanours such as interfering with an executor's right to bury his testator. Judges eventually allowed these last offences to develop into the new general crime of disinterring a corpse without authority. The first reported conviction for this occurred in 1788 when a body snatcher called Lynn was punished for entering a burial ground and removing a corpse for dissection. The court condemned his action as "highly indecent and against good behaviour." Initially these criminal offences applied more to the body snatchers than to their customers, with the absurd result that once a corpse left a body snatcher's hands its return could not be ensured. Although an anatomy school could have no right of property or ownership in it, neither could anyone else, and the only persons who had any kind of a claim at all were executors and close family members asserting not ownership but the right or duty of burial. Thus, when corpses were found in an anatomy school or elsewhere, there was no certain way of forcing their surrender unless they were first identified and then made the subject of a legal claim by the family or executor. The family possessed the right to have the body handed over for burial, no more.

By the third decade of the nineteenth century, public complaint was growing ever louder and more critical of the inaction of Parliament and the law. In response to this concern the House of Commons in 1828 appointed a Select Committee to report on the degree of social need to obtain bodies for anatomical examination. In May of the same year, the courts allowed the criminal law to reach into the anatomy schools themselves. At the Lancaster Assizes, Mr. Justice Bayley convicted and fined three defendants involved in the normal functioning of an anatomy school—a surgeon, a medical student, and an apprentice. They were found guilty of having "in their possession a body knowing it to be disinterred." The fines totalled £45, approximately 150 times more than the normal weekly wages of an agricultural labourer or an industrial worker. This trial had an electrifying effect on the medical profession and the anatomy schools, which in that year in London alone had 800 enrolled students. They began to join in the clamour for reform.

Retired and practising resurrectionists gave evidence in the hearings of the Select Committee. One of them disclosed that he and his gang had dug up and sold 1,211 adults and 179 smalls (children) in London in the five years 1809 to 1813, an average of around 280 per year. His "record" for rapid resurrections was twenty-three in four nights. Another witness said that his gang in 1809 and 1810 disposed of 305 adults and 44 smalls in England; 37 were sent to Edinburgh and 18 were kept "in hand and never used." A "small" was a body under three feet long. These were sold by the inch and were classified as "large smalls," "smalls," and "fetuses." The average price of an adult body around 1810 was four guineas, more than eleven times the typical weekly wage mentioned earlier. Teeth were also valuable and were normally removed and sold to dentists. Retainers and hush money were taken from the owners of anatomy schools at regular intervals.

Ben Crouch, one of the best known of the London resurrectionists, told the Select Committee that the use of armed guards on graveyards in London had made the trade much more hazardous and driven the price of a body as high as fourteen guineas. The grave robbers had inspired other measures of security, too, such as the placing of graves in locked metal enclosures and the use of iron coffins.

The report of the Select Committee, filled with alarming information and recommending reform of the law, was presented to Parliament in June 1828. In December 1828, after months of inaction, news from Edinburgh of the indictment and trial of a young, hard-drinking, bad-tempered, work-shy Irishman, William Burke, broke upon a horrified Britain. Burke and his confederates—William Hare (also an Irishman), Hare's wife, Mary, and Helen McDougal, Burke's woman—had made a total of £130 in the year between November 1827 and November 1828 by selling dead bodies to the private medical school of one Dr. Robert Knox.

It had all begun simply enough. The Hares conducted a lodging house in Edinburgh and lived on the premises. Burke was a lodger there. On November 29, 1827, another lodger, an old pensioner named Donald, died of natural causes owing £4 rent. Between them, Burke and Hare conceived the idea of selling Donald's body to Knox's medical school. To their surprise it was easy to do. No questions were asked and they were paid £7 10s. The retention of dead bodies by creditors as a means of forcing a deceased person's family to pay his debts was not unknown in those days. Burke and Hare were so stimulated by the easy money that they decided to make more by creating their own corpses. Their *modus operandi* was to befriend and invite lonely or derelict people to their rooms for a drink, and after a few glasses of whisky, to kill them. In this way they murdered sixteen of their guests, the last one an elderly widow, Mrs. Docherty, who was killed by Burke in his room during the night of November 1, 1828. He and McDougal had moved from Hare's to another lodging house, but Hare still worked with him in transporting and selling the bodies of the victims. On the morning of November 2, Mr. and Mrs. Hare came to Burke's room, where he had concealed Mrs. Docherty's body under a heap of straw. The four confederates then left the building, and while they were absent, an inquisitive visitor looked under the straw. The Hares were arrested on the same day, and Burke and McDougal on November 3.

To save his skin and his wife's, Hare agreed to confess and to give evidence against Burke in exchange for immunity from prosecution. Only Burke and McDougal were charged with murder. The trial began on Christmas Eve 1828, and because Scotland did not observe a holiday at Christmas in those days, continued on Christmas morning. The jury retired at 8:30 A.M. and returned in fifteen minutes with a verdict of guilty against Burke and "not proven" against McDougal. The judge immediately pronounced the death sentence in the following terms: "... to be hanged by the neck ... upon a gibbet until he be dead and his body thereafter to be delivered to Dr. Alexander Munro, Professor of Anatomy in the University of Edinburgh, to be by him publicly dissected and anatomized." The trial was all over by 10:00 A.M. and the execution fixed for the morning of January 28, 1829. A crowd variously estimated between twenty thousand and thirty-seven thousand watched Burke die on the gallows in the Edinburgh Lawnmarket. His body was immediately taken to the university where a huge and unruly mob besieged the Anatomy Theatre. The police were called to keep order. By the time Professor Munro had completed the dissection, the area of the classroom, according to one account, "had the appearance of a butcher's slaughter house, from its blood flowing down and being trodden on." The police were only able to quell the rioting by allowing people fifty at a time to enter the Anatomy Theatre and view the remains. Not less than twenty-five thousand gazed at what was left of Burke on that and the following day. According to reliable reports, Burke's body was cut up and pickled in brine, and a quantity of his skin tanned, sliced into portions, and sold. Some of it is said to have been made into a tobacco pouch, and the other pieces to have been printed with portraits of Burke and McDougal, and sold as curiosities. His skeleton was kept by the university and still hangs in a glass case in the museum of the School of Anatomy. To this day the word "burke" indicates "murder by strangulation; hushing up a matter."

As for Burke's former companions, one author says that Mrs. Hare left her husband and went to Belfast, where she disappeared from view, and that Helen McDougal went to Australia. A number of accounts record that Hare himself left Edinburgh and obtained work at a lime-burning works where he was recognized by his fellow workers and flung into a lime kiln and blinded. He then moved to London where "for a period of forty years he existed as a blind beggar moving about on the north side of Oxford Street."

As sensational as the Burke and Hare case was, it still did not inspire the Westminster Parliament to action. London was a long way from Edinburgh, and in March 1829 the House of Lords rejected a reform Anatomy Bill which had been passed by the House of Commons, after having been introduced there by Mr. Henry Warburton, M.P.

However, London was to have its turn. In November 1831, in New Kent Road near Lambeth, John Bishop and Thomas Williams murdered a boy of fourteen named Carlo Ferrari, or Ferreer. Carlo, who came from Italy, had made a living by exhibiting his two pet white mice, which he carried in a small cage slung around his neck on a string. Bishop and a third man, James May, offered the body for sale to the Dissecting Room at King's College in London and agreed to accept nine guineas for it. The anatomy demonstrator, Mr. Partridge, became suspicious when he saw a number of injuries around the head and observed that all the teeth had been extracted. It was later found that on the same day May had sold twelve human teeth to a dentist for one guinea. The dentist said that pieces of flesh and jaw adhered to some of the teeth, indicating they had been extracted with great violence. Partridge summoned the police, and the three men were arrested that afternoon when they came to collect their money. Bishop, Williams, and May were tried at the Old Bailey on December 1, found guilty of murder, and sentenced to death. Afterward, on December 4, Bishop and Williams confessed to Carlo Ferrari's murder and also to having murdered a woman and a boy for the same purpose. They exonerated May from involvement in the killings, and he was immediately reprieved. Bishop and Williams were publicly hanged outside Newgate Prison in front of thirty thousand angry spectators who later rioted. In accordance with the terms of the sentences, the bodies of the executed men were delivered to the Surgeons' Hall that evening "for dissection and anatomization." Within ten days of the hangings, Henry Warburton reintroduced his Anatomy Bill in the House of Commons. This time the Lords did not reject it and the bill passed both houses, becoming law on August 1, 1832.

The Anatomy Act of 1832 introduced the principle of licensing, which had long been known in Europe. Strict rules were imposed upon anatomy schools, including the licensing of both instructors and students of anatomy, supervision by government inspectors, and the filing of regular reports. A simple procedure was created whereby any person or his relatives could direct that his dead body be handed over for anatomical examination. Unclaimed bodies could also be handed over for the same purpose by those in lawful possession of them. After examination, the remains were to be decently interred. Imprisonment and fines were prescribed for offences. The act abolished a previous

law that had made it compulsory for the bodies of all executed mur-
derers to be either dissected or publicly hung in chains; the dissection
provision was replaced by one that gave the courts the option to order
either public hanging in chains or burial within the precincts of the
prison.

The Anatomy Act of 1832 was a simple and completely effective
piece of legislation that at one stroke destroyed the trade of the body
snatchers. Only now is its procedural method being set aside in some
places to make way for laws more suited to modern conditions.

Despite its drama, the bizarre story of the British body snatchers does
not prove that any widespread monetary value was attached to the
human body in former times. Bodies certainly commanded a considerable price for a restricted period in the vicinity of anatomy schools, but
this was really a black market operating outside the law or on the
fringes of the law. The demand was highly specialized and limited in
quantity.

Has the dead human body ever had any other kind of intrinsic
value? In some societies both ancient and modern, religious or spir-
itual beliefs have required that the body remain intact and undisturbed after death, and magical powers have been attributed to certain
organs, but there seems to be no evidence that the dead human body
has ever been generally accorded a special value. Strict Jewish religious law has traditionally required the burial of corpses intact on the
day of death and has forbidden mutilation of the body whether dead
or alive. Some Orthodox Jews still hold to these precepts, even to the
extent of disapproving of all forms of autopsy.

History has pointed to savage societies that have followed the practice of eating the organs of the dead—for example, the heart of a brave
enemy killed in battle—as a homeopathic way of absorbing his ad-
mired qualities. However, ritual cannibalism has not invariably been a
heroic affair nor has its occurrence been confined to antiquity. It is still
practised. A thoroughgoing analysis of the status of dead bodies in a
modern cannibalistic community is contained in a 1971 judgement of
the Supreme Court of Papua and New Guinea. In the course of finding that the consumption of cooked human flesh taken from a dead
body by members of a remote Papuan tribe was neither an improper

nor indecent interference with the body, the (European) judge cited instances in which consumption occurred "as a reverence or ritual," and others where the "only significance in the eating would be gastronomical" or "purely for fresh meat." Papuan customs were cited in which the heads of defeated enemies had been eaten so that the courage of the dead man "might be subsumed." In cases of both kinds, ritual importance was often attached to the manner in which the body was treated after death and before eating. A different attitude was displayed by Emperor Bokassa I of the Central African Empire, who was deposed in 1979. His successor as leader of the Empire was reported in the London *Daily Telegraph* as saying that human flesh "trussed up and ready for roasting" had been found in Bokassa's refrigerators after the *coup d'état*. By way of comment, President Senghor of Senegal made the memorable and entirely accurate remark that "Europeans simply don't understand cannibalism."

Some Christian attitudes—those critical of cremation of the dead, for example—have been based on the belief that the human body ought to be kept whole after death in order to be resurrected properly upon the Day of Judgement. This attitude is diminishing, however, particularly since the Roman Catholic Church modified its approach to cremation in 1963. Prior to that time, canon law prohibited cremation of corpses as a general rule (exceptions were allowed, for example, during wars and epidemics). The Church's proscription was based on the traditions of the Old Testament and on the example of Christ, whose body was consigned to a tomb; it also was intended as a response to the secularist view that cremation confirms that death is the termination of all existence and excludes the possibility of resurrection. But in 1963 the Church outlined its present discipline in *The Rite of Funerals*: "Christian funeral rites are permitted for those who choose to have their bodies cremated unless it is shown that they have acted for reasons contrary to Christian principles.... These funeral rites should be celebrated ... in a way that does not hide the Church's preference for the custom of burying the dead in a grave or a tomb, as the Lord himself willed to be buried."

As a means of disposing of the dead, cremation had been long practised before the rise of Christianity. It is an ancient custom in India and other Eastern countries, and was introduced to the Western world by the Greeks as early as 1000 B.C. All devout Hindus wish to be incinerated in Benares and their ashes deposited in the River Ganges. In Tibet cremation was traditionally the privilege of a favoured few, being reserved for the bodies of high Lamas. The original Greek adoption of cremation is thought to have resulted from the example of Northern societies as a means of enabling the families of fallen soldiers to conduct appropriate funeral rites. The bodies of soldiers killed on foreign soil were cremated on the spot and the ashes returned to their homeland so that a ceremonial entombment could take place.

All these events and beliefs, whether ancient, modern, religious, superstitious, animalistic, or venal, have some bearing on the question whether the dead body has possessed an intrinsic value in the past. Some of them attached to it a limited, sporadic value, either social or commercial, but they were all of a different order and quality from the factors that give it a value today. These are entirely corporeal and capable of being precisely measured. The true reason for the unique worth which the body now has is the newly discovered capacity of its tissues and organs to cure disease and repair defective bodies. Its value, then, is a function of its therapeutic utility, and one would be wrong to conclude that it is a consequence of demand and shortage. The gyrations of the international gold market have shown that a scarce commodity will find its true price in a free market. Human tissues would do the same, but they could never be simply commodities, and when talking of them, value must not be confused with price or the cause of value with its effect.

In the past three decades, the ever-growing ability of science and medicine to demonstrate the extraordinary curative powers of body tissues has made them precious. As one writer has put it, a man may now draw upon his own body "from the very marrow of his bones to the outer covering of his skin" for dedication to another. This has generated intense demand, acute shortage, and a variety of unsuccessful systems aimed at securing adequate supplies. In these circumstances, it is hardly surprising that some people are prepared to buy organs and tissues while others are prepared to sell them.

There are four main reasons for the enormous increase in the present demand for body parts. The endless requirements of medical research and experimentation are the first. A gynaecologist engaged in endocrine research requires "buckets of urine." A surgeon who developed a functioning metal kneecap first studied dozens of human kneecaps from the morgue. All tissues are in great demand by researchers. Every part of the body will evoke interest from a researcher somewhere, whether that part be from a living person, a corpse or an aborted fetus, or whether it be a necessary part of the body, as vital as the brain or the heart, or only waste like sweat, urine, or cut hair.

The second reason is the growing usefulness of human tissues for the preparation of therapeutic extracts to treat diseased and defective patients. Official programs throughout the world are calling for more and more tissues for this purpose, for example brains and placentas for extracts that will activate the blood-clotting system. Hearts, kidneys, bones, intestines, pancreases, skeletal muscles, prostates, lungs, livers, tongues, and skin are needed for the preparation of serum-based solutions and cell cultures. Only one of the many uses of fetal tissue is for the preparation of vaccines.

The third reason is provided by the traditional medical activities of anatomy and autopsy. The vital importance of bodies for medical education is well illustrated by the story of the British body snatchers. Anatomy schools the world over have always needed a continuous supply of human cadavers. The quality of any community's medical profession will be governed by the efficiency of this supply and the encouragement given to autopsy. Many medical practitioners regard autopsy as essential to maintaining high standards of medical knowledge, hospital care, and community health. It is of primary importance to know precisely why people die. The need for autopsy and anatomical examination has not diminished in recent years, but neither has it greatly increased. However, changing methods in the preparation of anatomical exhibits and other developments in both fields make it necessary for every society to look to its best interests and facilitate the donation of dead bodies and the conduct of postmortem examinations.

The fourth and far and away the largest single cause of the worldwide clamour for human body parts is the escalating demand for organs and other tissues for transplantation. Its requirements are probably greater than those of the other three activities combined. Surgical know-how and supporting technology are highly developed and improving all the time. The prospects of success of established transplant surgery are encouraging, ranging from reasonably high to almost certain, depending upon the body part and upon the tissue compatibility of the parties. The numbers of sick and dying patients in need of tissue will not decline in the foreseeable future.

Probably the best-known shortage is that of kidneys, but there are also severe shortages, according to time and place, of other vital organs such as the heart and liver, of glands like the pancreas, and of corneas, skin, bone marrow, and blood. The scarcity in the United States has been described by one ethicist as "a deadly scarcity" in a lengthy essay entitled "Our Shameful Waste of Human Tissue," and by another as simply "the crisis." In 1974, an American medicolegal publication suggested that the death rate in severe burn cases, which was as high as 40 percent for victims of extensive third-degree burns (he was using California statistics), could be drastically reduced "if enough human skin were routinely available." In 1979, *The Times* of London reported a claim by a leading British surgeon that "thousands of patients are suffering needlessly because of shortage of donors." In January 1980, the chairman of a British government Working Party, which had just produced a Code of Practice for the removal of cadaveric organs for transplantation, stated that there still remained "a great shortage of human tissue for transplant" in Britain.

Even with a familiar and regenerative tissue such as blood, scarcity can rapidly develop. In the Soviet Union, visiting Western experts have seen evidence of serious blood scarcity in the recent past. In the mid-1960s, blood for transfusion was being obtained from dead bodies to supplement the national supply. To encourage Russian citizens to give their blood, extraordinarily large payments were offered in many places, but even the average payment was high, being the per pint equivalent of two weeks' minimum wages.

It is convenient and instructive to use the 1950–1960 decade as a reference point. During that period, there was a coincidence of significant events. Improvements in intensive care techniques and in machines for maintaining circulation of the blood and respiration drew attention to the dividing line between life and death in patients with irreversible brain damage. In turn, the knowledge gained from these developments made it plain that transplantation of whole organs taken from a brain-dead person whose heart and blood circulation were kept going solely by machinery had a far higher prospect of success: the "fresher" the kidneys, liver, heart, et cetera, the better the prospects of survival for the recipient.

The first successful kidney transplant was performed in the United States in December 1954 at the Peter Bent Brigham Hospital in Boston, using a living donor. The donor and the recipient were twenty-three-year-old male twins. By 1967, the United States Surgeon-General estimated that some 8,000 patients annually needed a kidney transplant in the United States. In fact, in that year, no more than 450 kidneys were transplanted. By the mid-1970s, the number had risen to 2,000 a year, and the American Medical Association was projecting an annual need of 10,000–20,000 kidneys. By 1978, regional centres in the United States were complaining. The New York and New Jersey Regional Transplant Program found itself at a standstill because of lack of donors. *The New York Times,* in an article on February 5, 1978, said that the region's annual transplant rate was 300–400 kidneys, but the permanent waiting list had reached 600. Statistical information published by the Department of Health, Education and Welfare for the year ended December 31, 1978, indicated that some 4,000 kidneys were transplanted in that year in the United States.

In February 1975, the British Transplantation Society published in the *British Medical Journal* the results of a formal inquiry under the heading "The Shortage of Organs for Clinical Transplantation." The report opened with the words: "The four transplant surgeons present were in agreement that there was a grave shortage of kidneys for transplantation. . . . The shortage of donor kidneys blocked access to treatment of new patients with chronic renal failure." At that time Britain's annual kidney transplant rate was 450, with about 2,000 patients in need of kidneys. Exactly five years later, one of Britain's best-known kidney transplant surgeons was able to draw attention to the nation's "enormous shortage of kidneys" by disclosing that one of his patients had been waiting for seven years for a transplant. Kidney transplants were then running at around 900 annually. The permanent waiting list was standing at around 1,500, with over 500 having been on the list for over two years.

In 1976 the need for more kidneys for transplant in France (which had a rate of some 350 per year with a waiting list of over 1,800) was so great that the French government passed a revolutionary law which may have a profound influence on the entire Western world.

Satisfaction of the demand for body materials has therefore become a major concern of governments and medical practitioners. Even if demand grows in the next thirty years at a rate no greater than it has in the past thirty years, the intensity of the claims for access to human bodies for their contents will cause serious social problems. There is reason to believe that the demand will increase at a much greater rate than in the past.

Of course, transplantation surgery uses far more tissues than kidneys, and it did not start a mere thirty years ago. Transplantation of tissues from one person to another has been attempted since earliest history. Historically, skin was the first tissue transplanted. Woodruff, in his medical textbook *The Transplantation of Tissues and Organs,* says, "Skin flaps from the forehead and the cheeks were used by Indian surgeons more than 2,000 years ago, principally in the operation of rhinoplasty" (rebuilding noses). The fact that this kind of transplant was of the patient's own tissue does not prevent one from marvelling at the skill of surgeons practising before the beginning of the modern era. Thomas Rowlandson, the late-eighteenth- and early-nineteenth-century English artist, painted a picture of the transplantation of teeth in England.

For better understanding, medicine has now divided transplantation into three categories according to the type of tissue used. The transplant of a person's own tissue from one part of his body to another—for example, in a skin graft—is an "autograft" or "isograft." These expressions are also used on occasion to describe transplants between identical twins, whose tissues are biologically the same. A transplant from one person to another is a "homograft" or "allograft," while the transplant of animal tissue into a human is a "heterograft" or "zenograft." The words "prostheses" and "prosthetics" are used to describe the implant of entirely artificial materials into the human body.

It was not until the nineteenth century that the transplantation of tissues from one person to another began to be successful. It has been claimed that the first successful bone transplant took place in 1878, and that the first cornea "that remained clear" was grafted in 1905. From that time until World War II, transplantation was attempted only intermittently and was largely confined to skin grafting, principally because the body rejected "foreign" tissues. The processes of rejection only began to be understood during World War II.

Today, some twenty-five different kinds of human body parts are being transplanted with increasing frequency. These include parts of the inner ear, a variety of glands (pancreas, pituitary, thyroid, parathyroid, and adrenal), blood vessels, tendons, cartilage, muscles, testicles, ovaries, fallopian tubes, nerves, skin, fat, bone marrow, and blood. The word "transplant" has also been used to describe the surgical interchange of fingers and toes, as was done with a patient who lost all the fingers of one hand in an accident and received the graft of his big toe and one finger of his other hand in order to give the mutilated hand some manipulative function.

The modern era of transplantation began in the late 1940s, with medicine exhibiting increasing interest in cornea transplantation and obtaining society's approval to use the eyes of dead donors. With the advent of successful kidney transplants, the horizon broadened greatly. After the first such graft in 1954, one medical writer wrote with admirable understatement, "The rapid restoration of the patient's health to normal was a remarkable stimulus to laboratory and clinical study." The performance of kidney transplantation spread quickly around the world and is now commonplace on every continent. Some twenty-one years later (in May 1976), there were 301 transplant teams at work in the world, and over 25,000 transplants had been carried out. More than 10,300 recipients were alive with functioning transplanted kidneys, and the longest surviving kidney had been functioning in the recipient for more than twenty years. At the Seventh Congress of the international Transplantation Society in Rome in 1978, it was announced that the total world number of kidney transplants had then passed 40,000. Recipients of transplanted kidneys can now expect to lead normal lives in every way. Many resume, or commence, strenuous sports. Many have married and had children.

The first liver and lung transplants were performed in 1963 in the United States. The recipient of the liver died within three weeks. The transplant surgeon, Dr. Thomas E. Starzl of the University of Colorado, was able to report fourteen years later in the *World Journal of Surgery* that some 275 liver transplants had been performed worldwide in the intervening period, of which he and his colleagues had done 40 percent. The longest survival was nearly seven years.

Lung transplants have not been successful. In fourteen years after 1963, no more than forty lungs were transplanted throughout the world. The recipient of the first, a prisoner serving a life sentence for murder, died eighteen days after the operation. The longest survivor lived ten months. Lung transplantation has been described as no more than "exploratory."

In February 1969, the first transplant of a larynx was successfully performed in Belgium.

The first heart transplant to a human being was attempted in 1964 in Jackson, Mississippi, with a chimpanzee's heart. It failed. The first successful heart transplant, which was performed by Dr. Christiaan Barnard and accompanied by explosive international publicity, took place in December 1967 in South Africa; the recipient died from double pneumonia eighteen days afterward. Yet, a heart transplant performed in the United States only ten months later caused the recipient to survive for another eight years. When she died in 1977, at the age of fifty-seven, she was described as the "world's longest-surviving heart transplant patient." World numbers for heart transplants were then about thirty a year, with most being done in the United States. By 1980, one institution alone, the Stanford University Medical Center in California, under the surgical leadership of Dr. Norman Shumway, was performing more than half the world's heart transplants, showing a one-year survival of 69 percent and a five-year rate of 50 percent. This inspired other medical groups to develop new programs.

The first successful heart transplant in 1967 had a prodigious impact upon the Western world, to a large extent because of the historic mystical role of the heart as the seat of the emotions and the indicator of life and death. The medical profession and the public in Western nations were so affected by the feat of the South African surgeon, Dr. Barnard, that there was a rush to copy him. It has been reported that in the succeeding year, 1968, 101 heart transplants were done by 64 medical teams in 22 countries. Most of the recipients died within a few weeks. The failure rates were so high, and the surgeons so inexperienced, that a strong adverse reaction developed and many hospitals throughout the world gave up heart transplantation. Dr. Shumway has been reported as saying, "It wrecked the field for a good five years."

In the late 1970s, there was a global resurgence in heart transplants thanks to improved surgical and aftercare techniques. Dr. Barnard began to use a technique of "piggy-back" hearts, leaving the original, diseased heart in place and positioning the transplanted heart in the right side of the chest cavity. In November 1979, he said that he had performed twenty-four such transplants, his purpose being to provide a kind of insurance or "fallback" for the recipient in the event of transplant failure. With a normal heart transplant, failure obviously means instant death. There was a resumption of heart transplants in England in 1979 after a voluntary five-year moratorium by the medical profession; some thirty hearts were transplanted by December 1980 at the rate of approximately two per month. There are now two major heart-transplant units in England, one at Papworth near Cambridge, and the other in West London. Dr. Shumway's Stanford team was receiving some three hundred referrals annually by 1980; no more than 15 percent of these were found to be suitable transplant candidates, and of these at least one-third were dying of heart disease before a suitable heart could be found. Despite the improving medical results, the supply of hearts is still inadequate.

During 1979, "double" organ transplants were performed in England by the famous English surgeon Professor Roy Calne. In August of that year, at Cambridge, he transplanted a kidney and a pancreas into a woman who had severe diabetes complicated by kidney failure, and in October, he transplanted a liver and a pancreas into a young man suffering from terminal liver disease and diabetes.

It is hardly surprising that the transplantation of entire human organs still attracts newspaper headlines and engenders feelings of awe in most people. It still seems incredible that surgeons can take a dying man who is too weak to walk more than a few steps unassisted, remove his diseased heart, replace it with the heart of a dead man, and within thirty-six hours have the recipient spending a brief time out of bed, talking to his family, and eating a steak. Spectacular achievements of this kind tend to overshadow other, more familiar therapeutic miracles worked with human tissues. Some forms of transplantation have become so commonplace and such a normal part of modern living that they are barely seen as transplants at all, and certainly not as miracles. The most outstanding is blood transfusion. No activity surpasses it in lifesaving capacity, simplicity, and widespread routine practice. Understanding of the vital importance of blood is age old, but the ability to transfer it safely from one human being to another is a twentieth-century development, proceeding from a scientific discovery in 1901 that led to the identification of the four principal human blood groups, A, B, AB, and O. For many centuries, misunderstandings of the nature and function of blood supported the practice of bloodletting. Richard Titmuss, in his 1970 book *The Gift Relationship,* described the course of medical enthusiasm for this treatment in the following words: "The trickles of purged blood in ancient Rome had become rivers by the eighteenth century when many surgeons, and even psychiatrists, reduced their therapy to the maxim 'purge and bleed.' This practice was abandoned only in the twentieth century."

World War II was the starting point of an immense expansion in the transfusion of blood. Since that time the process has become an indispensable part of medical practice. Scientific advances continually increase the utility and therapeutic power of blood. The demand for blood is vast and unsatisfied. Nations that cannot get enough from their own citizens import it from abroad. It is the one human tissue that has forced otherwise disapproving governments to make exceptions to the prohibition or discouragement of commerce in body materials. Its lifesaving power is so great and immediate that democracies have passed laws allowing the administration of compulsory blood transfusions to children in the face of parental opposition whether based on religious or other objections. Although there is no scientific objection to the use of blood taken from dead bodies, it is normally taken from living donors. Human placentas are also frequently used as a source.

During the seminal decade of the 1950s, philosophers, moral theologians, and the churches began to debate the long-term implications of tissue transplantation. On November 24, 1957, Pope Pius XII, in an address to an International Congress of Anaesthesiologists in Rome, answered questions on the diagnosis of death in unconscious patients, one of the principal sources of organs for transplant. From then until the present day, increasing attention has been given in the press, on radio and television, and in symposiums, seminars, and academic journals to the moral and ethical ramifications of the use of the human body in transplantation, research, experimentation, and therapy.

Powerful objective confirmation of the new value assumed by the body in the Western world is available from two sources. The first is that its medical utility has forced all nations to change their understanding of human death. The second is the proliferation since 1950 of public laws dealing with the removal and use of body materials. Although the most useful tissues, apart from those of the living, come from a relatively small number of people (those whose dead bodies are healthy, in a hospital, and "fresh"), the community now expresses to all living citizens a strong interest in the use of their corpses. The reason for casting the net so wide is that the person who will be the best source of tissue will not be recognized until death is imminent and the body parts are physically available to perform their therapeutic function. Faced with such unprecedented circumstances, it is no wonder that nations have found it necessary to remodel existing laws or to promulgate new statutes which directly regulate the use of bodies for all or some of these purposes. Such lawmaking activities were not confined to Europe and North America, but were equally vigorous in many other parts of the world.

Some of the more advanced Western communities, in particular California and a few other American states, have broken new ground with laws aimed at protecting the human body against some undesirable effects of medical and scientific progress. These include statutes recognizing the concept of brain death, "natural death" Acts which give citizens the right to refuse medical treatment that might drag out the agony of dying by the employment of machinery or artificial processes, and laws which allow the discontinuance of treatment to longterm coma patients who are permanently unconscious but not dead.

Because the body is now subject to previously unimagined treatment, it is possible that society will be tempted to treat it as a commodity. We are by now familiar with the idea of the removal and use of many body parts from both the living and the dead. The means of familiarizing the public with this have often been spectacular. In 1977, a thirty-two-year-old American man was saved from death by the kidney of a dead Russian teenager. Russian doctors had sent the kidney by jet from Moscow to New York. On December 29 of the same year, *The Times* of London reported that Winston Churchill, a member of Parliament and grandson of the Prime Minister, had piloted his own aircraft from Gatwick to Hamburg through a force-nine gale to deliver the kidney of a dead man aged forty-four, which was then successfully transplanted to an eight-year-old West German boy. At Christmas a year later, the kidney of a dead New York youth was flown across the Atlantic by Concorde and transplanted to a young Englishman in Portsmouth. The recipient made a full recovery and in the following year married and became the father of a child.

Increasing familiarity with the curative powers of human tissues is likely, in the opinion of many, to foster a public attitude under which citizens will be considered to have a duty to make their dead bodies available for the aid of the sick, and the community will have valid claims upon its dead for the same purpose. It is well within contemplation that as medicine continues to find greater and greater uses for human tissues, we will come to see these claims upon the dead assuming the aspect of a public entitlement. If this happened, the human body would acquire some of the attitudes of property. The possibility of such a climate of opinion makes it necessary to reexamine the strength of our beliefs in personal autonomy and individual freedom.

The suggestion that lifesaving medical advances could imperil individual freedom may seem far-fetched, perhaps even melodramatic, but there is a more cogent reason for it than the mere growth of transplantation. The latest Western laws, particularly in Europe, indicate that a more peremptory social outlook has already developed, under which consent is no longer required before body parts may be removed from the dead. At the same time, we are experiencing an outburst of medical and scientific activity which could produce a discrete but equally novel classification of many human beings as objects or commodities. This relates to human reproduction and could feasibly presage the formulation of new concepts to control the custody, or even the ownership, of many children, the interchangeability of parental rights, and the use of human reproductive tissue for transplantation and research. Specifically, we are here talking of human artificial insemination, the "test-tube babies," the use of fetuses as sources of body parts for therapy and research, and the transplantation of reproductive tissues such as ovaries, fallopian tubes, and testicles.

Once again, the story of blood transfusion offers cause for reflection. Blood is a regenerative tissue, and this characteristic may have tended to obscure issues for personal autonomy which are more clearly suggested by the removal of tissues which cannot regenerate. The importance of the distinction is accepted by various transplant laws—for example, the model laws of the Council of Europe, Canada, and Australia, and the South African Act. The Australian, Canadian, and South African laws all use the expression "tissue that is replaceable by natural processes of repair." The capacity of regeneration is possessed by a number of tissues. Skin, bone marrow, semen, cartilage, tendon, bone, hair, and fingernail are all regenerative. There is an obvious difference between removing a small amount of blood for transfusion and removing a kidney for transplant, even if the donor is in the best of health. Legal recognition of this difference by provisions of differing severity in a statute is clearly justifiable. A lawmaker should also be aware that within each of the two classes of regenerative and nonregenerative tissues there are varying degrees of seriousness for transplantation. The removal of skin is a very painful experience and some surgeons refuse for that reason to take it from living donors. Removal of bone marrow and some other regenerative tissues, while not so painful, must be done under a general anaesthetic. By contrast, a pint of blood can be taken quickly and with little inconvenience and will be replaced by the donor's body in about six weeks.

The simplicity and safety of blood removal on the one hand, and its dramatic capacity to prevent death on the other, may well be responsible for the failure of many nations to appreciate that it is not in a category of its own but is very much part of the emerging pattern formed by curative, transplantable body materials. Writing in 1970, Titmuss cited twenty-seven nations around the world in which payment for blood was normal and five in which substantial fringe benefits such as free holidays were given in exchange. Nine of those in which blood was bought for money are members of the Council of Europe. Of those, Sweden was unique in paying cash for all its blood. This information does not sit comfortably with the words of the report of the Council of Europe of April 1975: "Almost universally the general principles of civil law in Member States consider illegal any transaction for selling and purchasing of this type of biological substance." It may be that the report intended to exclude blood from this comment on commerce, but it did not say so. Further indications of the deep-rooted, unself-conscious acceptance by many that blood is entitled to occupy an anomalous position was offered by South Africa in 1970 and Canada in 1971 when their new statutes on body-part removal prohibited commerce in all tissues *except* blood. Nevertheless, when the Council of Europe produced its model transplant code in 1978, which specifically included blood in its reach, it provided unhesitatingly that *all* forms of profit-making from the disposal of human tissues should be illegal.

One by one, activities have been tolerated in the quest for blood that arguably would not have been accepted with other body parts. Widespread commerce, compulsory transfusion, and unregulated removal are all well-known, although not all are frequently found in a single statute. Still, blood transfusion, standing alone, would not necessarily invite special attention from those interested in the preservation of individual freedom. However, when one considers the many medical and scientific demands now made upon the body for other tissues, a different opinion is justified.

To predict the appearance of overriding legal rights whereby individuals or the state or both would be entitled to take and use body parts might strike some people as absurd. Yet there is a depressing historic precedent for the treatment of the human body as property. The precedent is slavery, an institution whose destruction in the West required more than seven hundred years of struggle. It should be hastily said that no valid comparison can be drawn between slavery and the practices that are the subject of this book. Indeed, such a comparison is absurd. Slavery is inspired by man's greed and cruelty, while transplantation and other therapies that employ human tissues are designed for man's benefit. The purpose of mentioning slavery here is to show that it is not difficult to apply property concepts to the human body.

Slavery is the classic model of the living human body as property, and the classic form of slavery, its most pernicious, is chattel slavery. It has been known in the most backward and the most advanced ancient and modern societies—in classical Greece, imperial Rome, biblical Israel, and more recently, Japan, China, India, North and South America, Russia, Arabia, and Europe. Under it, slaves and their children belong to other men and may be treated as chattels, according to the owner's wish. Other forms of slavery include serfdom, sale and inheritance of wives, sale of children, debt bondage, caste systems, and forced labour in the style of Soviet Russia. Although all have been banished from the Western democracies, they have not all disappeared from the Eastern world, nor from the African and South American continents.

The two great international proclamations of the twentieth century on slavery are the Slavery Convention of 1926, and the Supplementary Convention on the Abolition of Slavery, the Slave Trade, and Institutions and Practices Similar to Slavery of 1956. They define a slave as a "person over whom any or all of the powers attaching to the right of ownership are exercised. . . ."

Sophisticated societies that have accepted slavery have, typically, created codes of laws to regulate it. The Anglo-Saxons had a fully developed, legally regulated system of slavery when the Normans conquered them in A.D. 1066. Slaves were not to disappear entirely from England itself until the eighteenth century. We should therefore be particularly sensitive to any claim for a measure of control over the human body and to any impingement upon personal autonomy, whether the object is the medical service of the community or anything else. One of the most instructive features of slavery in this regard was its institutional character and the respectability conferred by the legal rules that governed it. The profusion of laws regulating contracts to purchase and sell, title deeds to, and inheritance of slaves blurred the sharp edges of inhumanity. "After all, if the Parliament accepts it, and the community too, could it really be so bad?" The lesson for those concerned in facilitating and regulating the acquisition of human tissues for the benefit of the sick is that under slavery the human body became and remained property with disturbing ease.

With the complete abolition of slavery in the nineteenth century and the rise of the Human Rights Movement in the mid-twentieth century, Western man has banished both the possibility and the principle that one human being can own another. It would be ironic if in the aftermath of such events he created, without a full comprehension of their ramifications, new relationships in which a segment of society is given entrenched rights and powers, buttressed by the law, over healthy living persons and the dead as well, for the purpose of obtaining human spare parts.

In summary, the extraordinary utility of the body, the noticeable change in the underlying philosophy of the latest Western laws regulating the acquisition of body materials, the certainty of more medical advances, and the public's acceptance of the new concept of brain death and its increasing interest in gaining access to human bodies afford the best of reasons for a review of medical events of the past thirty years and an examination of society's response to them. One of society's demonstrable objects has been to produce a supply of human tissue sufficient to match demand. Has this object been achieved? If not, is the community now likely to react with stronger laws or procedures that put the individual citizen at risk? Subsequent chapters will address these questions and examine the options for action available now—but not necessarily in the future. Where are we heading? What ought we to do?

CHAPTER 2

The Demand for Spare Parts

I.

Before the response of society to the new utility of body materials for the cure of the sick is examined, in particular the "human spare parts" laws of the West, some fundamental questions should be asked and answered. The first is whether the spare parts themselves are truly useful and beneficial. The next is whether there are any competing medical advances that might reverse the growing demand for human tissues. Two particular fields of interest are the use of animal tissues and the building of entirely artificial body parts. A converse question is whether any parallel developments have the potential, not of reducing demand for human tissues, but of accelerating it. Finally, there is a politically and emotionally volatile question: What influence should economic considerations be allowed to exert? Are transplants and all the other uses to which our bodily bits and pieces can be put worth the expense? Is it really desirable for quantities of blood, skin, organs, and other tissues to be taken from the dead and the living for transplantation, for conversion into therapeutic extracts, and for education, experiment, and research? The answer is a clear "yes."

Probably the oldest medical activities that use bodies are anatomical examination and postmortem examination (autopsy). Both are integral to medical education and both necessitate the removal and examination of body contents. So important is autopsy regarded as a means of ensuring excellence in the quality of medical care that many hospitals in the United States and elsewhere insist on maintaining an autopsy rate of at least 90 percent of all deaths within their walls.

Of all body materials, blood is the most widely used and most easily obtained. The simplicity of its removal and transfusion can obscure the fact that the process is a true human-tissue transplant. Nobody knows how many lives blood transfusion saves annually, but just as oil is a vital liquid for modern industrial society, blood is a vital liquid for modern man. Its usage in the treatment of the sick has increased at a tremendous rate. In the thirty years between 1946 and 1976, English blood collections multiplied more than seven times from (in round figures) 200,000 pints to 1.4 million pints annually. In the United States, from the time when national estimates began to emerge, in 1956, collections more than doubled to beyond 10 million pints in 1979.

As for surgical transplantation of body parts, including whole organs, one would not have thought it necessary to argue its benefits. A visit to the kidney dialysis ward of any hospital or a talk with the recipient of a successful kidney transplant will prove the point beyond all doubt. How can a comparison be made between a healthy, energetic person leading a normal life, and the person he was, continually sick and spending six hours at a time three days a week in a hospital attached to a machine through permanent openings in his wrist veins?

The following questions and answers were asked and given in a public inquiry into transplantation in Australia on March 23, 1977:

Q. You are a patient who has had a successful kidney transplant?
A. Yes.
Q. Tell me of the difference in the quality of life of a man on dialysis and a man with a transplant.
A. Well, I regard life on the machines as just existing and not living, really ... I was very ill. ... I really could not enjoy a social life because I could not partake of the normal things that people did.
Q. Tell us something about your life since the operation.
A. For the last five years I have led a very normal life. I can eat or drink anything I want to.

Perhaps the most vivid example of the difference that a transplant can make is an extraordinary athletics contest that took place in Southsea, near Portsmouth, England, in August 1978. Competitors, every one of whom had received a successful kidney transplant, came from all over the world to take part in this first "Transplant Olympics." Its purpose was to show the world that a patient who has received a transplant can be totally cured and lead the same kind of life as anyone else. Two of the contestants, a young married couple, had received three kidney transplants, the wife two and the husband one. They won three bronze medals, the wife one for tennis and the husband two for sprinting and squash. In 1979, they made medical history when the wife gave birth to a healthy baby boy in London. So successful were the first Transplant Olympics that they were repeated in 1979 with teams of competitors from fourteen nations. The athlete who lit the 1979 Olympic flame, a popular English television entertainer named Jimmy Savile, described the contest as "a modern miracle," drawing attention to the fact that forty years earlier, before the invention of the artificial kidney machine and the technique of viable transplants, such an occasion would have been impossible because all the participants would have been dead or dying.

The transplant surgeons themselves feel some of the pleasure that is created by successful transplants. In the words of a well-known American practitioner, "It is a constant joy and thrill for the attending surgeon to see a chronically ill, debilitated patient, often full of pain from his fingertips to his toes with swollen abdomen, weakness, anaemia, bleeding, and fetid breath, restored to normal within days after the operation."

Kidney transplant has now reached the stage where, even with the problems of bodily rejection, high success rates are normal. It can be expected that 60 to 70 percent of transplanted kidneys, whether taken from dead or living donors, will be functioning a year after the operation. When the donor and the recipient are close blood relatives, the success rates mount until with identical twins they reach 100 percent. In September 1977, a British study group demonstrated that the prospects of survival for their kidney transplant recipients were approximately the same as for those patients who stayed on the kidney dialysis machine, while their quality of life was far superior.

Perhaps most important of all are the words of Sir Peter Medawar, who was awarded the 1960 Nobel Prize for Medicine and whose work on skin grafting and the body's immune-rejection mechanism is in large measure responsible for the rise of modern transplantation. Speaking to a Transplantation Congress in New York in 1969, he said, "The transplantation of organs will be assimilated into ordinary clinical practice . . . and there is no need to be philosophical about it. This will come about for the single and sufficient reason that people are so constituted that they would rather be alive than dead." Another of his publicly expressed opinions is that both hearts and livers will become as widely transplanted as kidneys.

To demonstrate the worth of human tissues in forms of medical treatment other than transplantation, one can cite both well-established and developing programs that all need greater and greater supplies of body materials. Among the most striking illustrations of public benefits are national programs involving the pituitary gland. In the decade 1967–1977, under an Australian national pituitary program (which maintains close contact with the U.S. National Pituitary Agency), approximately 300 children who would otherwise have remained dwarfs were successfully treated with pituitary hormone extracts, and some 250 previously infertile women were enabled to bear children after treatment with a different extract made from the same gland.

The benefits to be gained from supplying human tissues for medical education, legitimate research, and experiment are self-evident.

It is safe to say, then, that the use of body parts in medicine and science has been deemed acceptable in modern society. Nationally and internationally, transplant societies and associations abound, as do governmental, charitable, and purely private programs whose sole purpose is to increase the supply. More and more governments now routinely send organ-donation documents out with driver's license renewals. One of the latest proposals, devised by the West German government, is to provide, on the identity card that every citizen must carry, an indication of his wishes for the use of his body after death.

So that readers may make their own judgements on the worth of body-part donation, a fictitious case history is now offered without comment. It concerns ordinary people in everyday circumstances, and accurately presents the kind of events that typify cadaveric kidney transplants. This case is based on the English *modus operandi,* but similar results could be expected in most Western nations.

Gilbert discovered that he had kidney disease shortly before his thirty-seventh birthday. He was a schoolteacher, married, with three children under the age of ten. During a routine medical checkup, Gilbert's doctor found that he had high blood pressure and protein in his urine. Gilbert still felt perfectly well, but further tests showed that he had only 50 percent of normal kidney function and was suffering from inflammation of the kidney filters. He had a form of glomerulonephritis. After two more years, his kidney function was down to 10 percent of normal, despite medication and other treatment. He no longer felt robust, became exhausted very easily, and had a sickly complexion. His doctors decided that in order to stay alive he must begin treatment by hemodialysis. This first involved an operation on his left arm to create a fistula—a connection whereby an artery is joined to a vein and the skin closed over the join. A fistula causes the vein to become enlarged so that needles can easily be inserted to connect the patient to a hemodialysis machine (artificial kidney). Gilbert was given a controlled diet with severe restrictions on his daily intake of fluids, protein, and salt, and put on a routine of three dialysis sessions a week, each of five to six hours' duration. During each dialysis session, all his blood was routed through the machine fourteen times via tubes, filtered or "washed," and returned in the same way. By this means the blood's chemical balance was maintained, a job his kidneys were no longer able to do for him. Although Gilbert's health improved and he was able to resume some teaching, he was permanently sick and unable to take holidays or travel any substantial distance from the hospital. He then acquired a home dialysis machine, which meant that he could carry out his own treatment at home on three nights a week. This was an advantage over hospital attendance, but he was still the "prisoner" of the machine. To take matters worse, Gilbert began to have trouble with his veins after a year of dialysis. It became difficult to effect entries into them. Over the next two years, another fistula was made in his right arm and it became necessary to resort to veins in his legs and his groin to maintain access points for the machine's connections. About ten operations under anaesthetic were performed for this purpose.

During this period, Gilbert's doctors advised him to consider a kidney transplant. A successful transplant would release him from the bondage of dialysis and from the risk of reaching a point when his veins would no longer accept the connections, and enable him to return to full health and normal living. The disadvantages of transplantation included the normal risks of all surgery, a one-third risk of failure of the operation within twelve months, the necessity for lengthy immunosuppressive medication to forestall rejection, and the enormous emotional and mental strains of waiting for a kidney with no assurance that one would become available within any given time, or at all, and possibly having to cope with a failure and return to dialysis. Gilbert was quite aware that kidney patients die at all stages of the disease: on dialysis, under surgery, after surgery, and after failure of a transplant. Some deliberately bring about their own deaths.

Two years after starting dialysis, he decided to seek a transplant. His blood was tested and he was tissue-typed. Details of his tissue type were sent to the National Transplant Service and recorded both in normal files and on a computer. The computerized information is amended and checked twice a week and a monthly printout is sent to all registered transplant centres. In theory, a recipient's body is more likely to accept a kidney from a person with a similar tissue type and blood group. The service not only classified Gilbert by tissue type and blood group but by inclusion in one of five categories as well. These categories show variations in urgency of need and readiness of recipients to take the risk of accepting a kidney of lesser tissue compatibility. Some renal physicians believe that tissue typing makes no more than a 10 to 15 percent difference in prospective success, so that if the best tissue match from an unrelated (nonfamily) donor will confer approximately a 65 percent chance of success, the worst will still give a 50 to 55 percent chance. Gilbert agreed to be put in the middle category. He had to discipline himself to be instantly available at any hour of the day or night without having any idea when the hospital telephone call might come.

Gilbert had been on the National Transplant Service waiting list for a year, and on dialysis for three years, when an unconscious nineteen-year-old named John was admitted to the emergency room of a public hospital three hundred miles away. It was a Saturday night, and he had been a passenger in the front seat of a car which had collided head-on with a taxi. John was a student, and the driver was his boy-

hood friend and neighbour, Charles. They were on their way to a discotheque. The car was a total wreck and Charles had a number of
broken bones but nothing worse. John, who was not wearing a seat
belt, had sustained a tremendous blow to the skull when his head hit
the windshield. He had virtually no other injuries. In the hospital, he
was taken to the intensive care unit, and within a short time connected
to a ventilator, a machine that artificially forces air into the lungs
through tubes via the mouth, and sustains respiration, heartbeat, and
circulation of the blood. John never regained consciousness. Within
thirty-six hours, the hospital specialists knew that his brain had been
destroyed by the force of the impact and by deterioration and collapse
after his admission to the hospital. They spoke to John's parents at
length and informed them that they believed that their son was dead
but that they intended to repeat over the next twelve hours a series of
medical tests whereby the diagnosis could be confirmed. They asked
whether, if confirmation was obtained, the parents would agree to the
removal of John's kidneys for transplant. Feeling that such a gift
would do some good and make a little sense out of the tragedy, John's
parents agreed.

Medical ethics require that all tests and treatment carried out on a
patient before death must be for his benefit. However, when blood is
being taken for normal tests it is within ethical principles to take extra
blood for tissue typing. This was done in John's case, and the results,
together with other relevant details, were telephoned to the National
Transplant Service, where they were immediately fed into a computer,
which produced a list of the ten best-matched waiting recipients in the
hospital where John was a patient, and the twenty best-matched waiting recipients nationwide.

Gilbert's name was at the top of the nationwide list, while none of
the recipients in the list at John's hospital presented nearly so good a
tissue match. The surgeons there decided that they would not use
either kidney. They allowed the National Transplant Service to telephone Gilbert's hospital, where the offer of a kidney was promptly accepted. Gilbert received his telephone summons. In the meantime, the
coroner's permission for the kidney removals had been obtained.

Twelve hours after the interview with his parents, John was declared dead because of the total loss of his brain function. With their

knowledge, his body was taken into the operating theatre where the kidneys were rapidly removed. In order to ensure maximum prospects of successful transplantation, kidneys should be functioning virtually to the moment of removal. The longer the lapse of time between cessation of function and removal, the less chance they will have of working in the recipient. After John's death was pronounced, the ventilator was kept going, and this enabled the kidneys to continue to work. At the time of removal, they were fully oxygenated and "perfused," meaning that they could be immediately chilled and packed in ice with the full expectation of proper function if transplanted within two days. With the heart and the liver, this period is measured only in hours.

After the kidneys were removed, the ventilator had been switched off, and John's body taken to the hospital morgue as the first step in the process of interment.

While the kidneys were being removed, a radio-controlled pick-up vehicle was on its way to the hospital. The driver received the kidney designated for Gilbert in its small "ice box," and began his road trip. All transportation had been arranged by the National Transplant Service—which could also have used, under its established procedures, air transport by volunteer pilots registered with the Ambulance Association, regular airlines, or the railways.

When the kidney was delivered, it was accompanied by a sample of blood, a small quantity of spleen, and a lymph node, all taken from John's body for more exact tissue typing.

In the meantime, Gilbert had been able to go to his hospital without rushing. One advantage of his long wait for a kidney was that he had developed a complete familiarity with dialysis and an ability to face a return to the machine in the event of failure of the transplant. At the hospital he underwent a medical check and a short dialysis. Six hours after arrival he was taken to the operating theatre. The new kidney was surgically placed in his body. Drug treatment to suppress rejection was promptly begun. The operation itself was successful, although the new kidney did not work immediately, which is not unusual. For about two weeks, Gilbert continued to undergo dialysis, and then gradually he began to pass more and more urine as the kidney began to function. He was discharged from the hospital three weeks after the operation but was required to come back frequently for blood tests and examinations to detect possible complications. On a few occasions he was called back to have extra drug treatment for suspected rejection.

After three months, Gilbert was again teaching full time and attending the hospital less frequently. Four years later he has developed no signs of trouble. He still visits the hospital every four to six weeks and takes medication to inhibit rejection. This makes him more liable to infection, and he must take steps to avoid colds and the like. His diet is normal, and he is required to keep reasonably fit and to watch his weight. He feels as he used to before he became sick.

It is well to recall that more is included in the scope of the expression "human tissues" than just the organs of living and dead citizens—for one, the fetus and fetal material, tissues that occupy an important place both in the practice of medicine and in research. The British Peel Report of 1972 defined the fetus as "the human embryo from conception to delivery," fetal tissue as "a part or organ of the fetus, e.g. the lungs or liver," and fetal material as "any or all of the contents of the uterus resulting from pregnancy excluding the fetus, i.e. placenta, fluids and membranes." Fetal thymus glands are transplanted into babies and children with thymus deficiency so as to enable the recipients' immune-rejection system to operate normally. Placental tissue, too, has been found to have uses as a sterile dressing for severely burned patients, for example. In 1976, an American medical report showed that veins taken from umbilical cords of newborn babies, which vary in length from eighteen inches to four feet, can be therapeutically implanted in blocked adult leg arteries to forestall gangrene and amputation. Consideration was also being given to the use of such blood vessels for bypassing obstructed heart vessels and arteries in the brain. Fetal research was instrumental in producing poliomyelitis vaccine. The Peel Report listed fifty-one different kinds of research projects that utilized fetal tissue and materials. Some of this was indispensable, for example, for research into the cause and cure of the common cold. Not only is there no practical substitute for fetal tissues in areas of preventive medicine research but, in the words of the Peel Report, their use "has gone beyond research into the field of established practice."

The report also indicated their indispensability in cancer research, arterial degenerative disease, and congenital deformities. In 1972, *The Sunday Times* of London reported the transplant in Beirut of the testicles of a six-and-one-half-month fetus to a twenty-eight-year-old man. The fetus had died in its mother's womb and its testicles were transplanted to the inside of a thigh of the recipient, who had been born without testes. However, this kind of transplant was seen by British doctors at the time as having largely a curiosity value, because the same result could be achieved by conventional hormone treatment.

The acquisition and use of fetal materials require considerable sensitivity, particularly since their source is mainly abortions. The delicate balance between powerful moral, ethical, religious, and political considerations and medical needs has, up to now, been well-maintained not only in Britain but also in the United States, where in May 1975 the National Commission for the Protection of Human Subjects of Biomedical and Behavioral Research reported to the government on fetal research and its proper regulation. The report was not unlike the Peel Report and rested upon a similar philosophy. Later in 1975, new regulations promulgated by the Department of Health, Education and Welfare permitted the use of fetuses and fetal material in research, but under strict controls.

The Peel Report showed its own prescience in the following words: "During our discussions we have been constantly aware of the public concern and of the ethical problems surrounding the use of fetuses, fetal tissues and fetal material for research. . . . In general we feel that the contribution to the health and welfare of the entire population is of such importance that the development of research of this kind should continue subject to adequate and clearly defined safeguards." The Peel group, in what has become a recognizably British approach, opposed the introduction of legislation and regulations. Instead it recommended a Code of Practice without legal force. The code, like the American regulations, allows the use of fetuses and fetal material in research, with clear controls, including a prohibition of "monetary exchange."

It is almost impossible to obtain accurate statistics on the quantity of fetal tissue employed in public health projects, but it is known to be considerable. Because these kinds of projects have been legitimized by the Peel Report and the American 1975 regulations, and appeal both

to researchers and to transplant surgeons, the demand for fetal tissue
will increase.

II.

Human tissue is not the only material with which "spare parts" surgery is performed. Two other kinds of substances have been employed
for repairing and replacing defective, diseased, or destroyed body
parts: animal tissue; and material of neither human nor animal origin,
such as wood, metal, mineral, and plastic. Both are used extensively by
the medical profession, to such a degree that if Shakespeare had lived
in the transplantation era, he might, when writing *Hamlet,* have felt
obliged to add more detail to his resonant hymn to the human body.
"What a piece of work is a man!" would be startlingly accurate when
applied to a person with a metal hip or kneecap, a bionic arm, ox bone
in his leg, and part of a pig's heart in his chest cavity. To obtain a balanced view of the transplantation of human tissues, it is well to remember that medicine also continues to advance in these related
fields.

As we have seen, the first heart transplanted to a human being be-
longed to an adult chimpanzee. The recipient was a sixty-eight-year-
old man dying of heart disease, who survived the operation by only an
hour or two; the operation took place in 1964 at the University of Mississippi. In June 1977, Dr. Christiaan Barnard, the South African surgeon who had performed the first successful human-to-human heart
transplant ten years earlier, grafted a baboon's heart to a twenty-five-
year-old woman in Cape Town. Four months later, he grafted a chimpanzee's heart to a fifty-nine-year-old man. Both transplants failed
and the recipients died, the first within five hours and the second
within four days. The first animal-to-human kidney transplants took
place in 1963 and 1964 in the United States with kidneys from chimpanzees, baboons, and monkeys. Most failed quickly, but one recipient
survived with a chimpanzee's kidney for over nine months.

Ox bone and pig and ape heart valves have been successfully transplanted to humans. In England, the main source of skin for the treatment of severe burns is the pig. This usage of skin is of a different kind
from grafts of a patient's own skin. Grafts are used for small burned

areas and will actually become part of the surrounding tissue, growing over and healing the wound. Supplied from both the United States and English medical companies, the pig skin is sterilized and punched with minute pinholes; it has the feel of fine, dry, cream parchment. After it is soaked and softened for twenty-five minutes, it is laid on the surface of the burned area and acts as a dressing that enables the underlying wound to heal in a healthy environment. Many medical specialists use pig skin in preference to human skin, because the removal of skin for use by burn patients is intensely painful for the donor. Some English doctors have gone so far as to suggest that removal of skin from living donors is unethical except in the rarest of cases.

Animal tissues have been widely used in transplant surgery, but the extent of their usage is not comparable with that of human tissues, although animal tissues have some obvious advantages as a source of supply. There have been occasional calls at medical congresses for deliberate selective breeding of animals for organ transplants, particularly the higher primates. When Dr. Barnard transplanted the baboon heart in June 1977, he did so in response to an emergency that developed during a heart-valve operation. Reports indicated that he had been keeping two healthy baboons isolated in the laboratory for just such an occasion.

Man has practised the domestication of livestock and the control and breeding of animal prey for his own consumption for more than ten thousand years. In principle, the introduction of similar practices aimed at producing organs and tissues for therapeutic purposes should be acceptable. Whether this would be so in fact is another matter. It is possible that there could be resistance not only from animal lovers but also from those who believe that animal organs should not be implanted into the human body, no matter how scientifically justifiable the operation may be.

Medically, the use of live animals means that immunosuppressive measures can be planned and carried out upon the recipient before transplant so as to suit his best interests. Also, the organ can be taken when required. With human cadaver donors it can never be predicted accurately when an organ will be available, and thorough preparation of the recipient is not feasible.

In most parts of the world, the source of organs for transplant is normally the cadaver donor. An exception exists with kidney transplants in the United States, where a large proportion of kidneys has traditionally come from living donors. Because people do not die to suit the convenience of others, a state of affairs has come about with organ transplantation in which, in effect, the tail wags the dog. Recipients, transplant teams, and hospitals have to organize themselves to respond with speed to the sudden availability of organs. When an organ becomes available, everybody must jump to it. Thorough medical preparation of the recipient is usually not possible, and this may affect the prospects of success. In addition, the expense of transplant surgery must remain high while these conditions apply. From the standpoint of the sick and of the community, it would be far preferable if the precise time of availability of an organ were known, so that the most suitable, convenient, and economic preparation for surgery could take place, and the highest chances of success ensured. In today's world, which lacks both herds of animals raised specially for transplantable tissues and organs and queues of living and cadaver donors, this remains an ideal objective.*

As remarkable as heterografts are, they lack the spectacular science fiction quality that is beginning to surround prosthetic implants. Some experts take the view that prosthetic surgery will increase a great deal more than surgery involving transplantation of human tissue. According to a French medical historian, the oldest known prosthesis is a T-shaped crutch represented on an Egyptian fresco of the fifteenth century B.C. In the Middle Ages and during the Renaissance, artificial limbs began to appear, such as wooden legs and heavy metal hands with mechanical fingers. There was a time when every retired English seafarer seemed to have either a wooden peg leg or a hook where one of his hands or forearms had been. Even today many of us have an un-

---

* On December 11, 1980, *The Times* of London carried a report from Peking to the effect that Chinese authorities were considering the resumption of partly successful research on cross-breeding human beings with chimpanzees. A Shanghai newspaper had reported on December 9 that in 1967 scientists successfully inseminated a female chimpanzee with human sperm, in the north-eastern city of Shenyang. One of the researchers stated that a possible use envisaged for the progeny (which was to be classed as an animal) was to provide organs "as substitutes for human or artificial organs in transplant cases." The pregnant chimpanzee had died, he said, after Red Guards smashed the laboratory in which it was kept.

natural but distinct penchant to repair and replace defective body parts, e.g. teeth, with precious metals. We already have in practical use entirely artificial heart valves, heart pacemakers, blood vessels, hip joints, knees, articulated fingers, ankles, shoulder joints, arms and legs of different kinds, kidney machines, heart-lung machines, and such established veterans as contact lenses, hearing aids, and false teeth. Long-standing projects in the United States, Britain, France, and Switzerland aim to produce artificial blood, skin, bladders, urethras, tracheas, and limbs with the capacity of "natural" function.* In France, the implanting of electrodes deep in the brain as a means of controlling pain has been practised since 1967. Underlying all of this is a vast infrastructure of research into "bionic" machines and "biomaterials" that can be tolerated inside the body. This work has resulted in prostheses containing polyurethane (for example, for artificial breasts) and other plastics, aluminum, steel, ceramics, titanium, rubber, carbon, and phosphate. Very large sums of money are devoted to this area of human therapy.

A respected institution which has made great progress with prosthetics is the Division of Artificial Organs at the University of Utah in the United States. The director of the division is Professor Willem Kolff, a Dutch-born medical pioneer who has been described as the inventor of the artificial kidney. Professor Kolff qualified as an M.D. in Holland in 1938, and began building the first artificial kidney dialysis machine in 1940, during the German occupation. In 1949, not long before moving to the United States, he began work on a heart-lung machine, and in 1957 on artificial hearts. Under Professor Kolff and

---

* In June 1979, the *Journal of the American Medical Association* carried a report indicating "remarkable success" in Houston, Texas, with an artificial device that enables impotent males to achieve an erection. One writer describes the device as consisting of two small cylinders implanted on each side of the penis. A tiny tube connects them to a small reservoir of fluid, a pump, and a release valve. The person achieves an erection by pumping the fluid into the cylinder and reduces it by releasing the valve. It is "comfortably implanted and easily manipulated." According to a urologist involved in the development of this prosthesis, it had functioned satisfactorily in 234 out of 245 patients after a small number of early failures.

A British medical spokesman said that the Houston technique might prove too expensive for the British National Health Service, which had developed a cheaper prosthesis—"a sort of splint which achieves a semi-permanent erection and is a reasonably practical solution."

his colleagues, the Artificial Organ Program at the University of Utah has developed many devices. A brief description of four of the most remarkable will provide a glimpse of the extraordinary future that awaits the human body upon their perfection.

For certain kidney patients, the dialysis machine is a lifesaving device. However, while it does its job, the patient reclines in a chair or on a bed and is literally tied to the machine by the connections to his body. A regulation dialysis machine is about the size of a large domestic washing machine, but the University of Utah program has produced a Wearable Artificial Kidney, called WAK, that weighs about 8 pounds (3.6 kg) and can be strapped across the chest. During a session using the WAK, the patient is freed from the bondage of immobility. He is able to move about for fifteen or twenty minutes at a time, although he must then resort to supplementary use of a tank containing 20 litres of dialysing fluid for additional filtration of body wastes. The WAK with its supplementary tank allows the patient a measure of freedom that is impossible in a hospital. He can travel and take holidays. The Utah people suggest that the WAK should be used six days a week for three hours a day, as contrasted to the usual three days for six hours with the hospital machines. In every three-hour session, the patient will be mobile for up to one and a half hours. Another portable kidney machine, capable of being packed in a suitcase with two weeks' supply of dialysing fluid, has been developed in England.

One of the people responsible for WAK, Dr. Robert Stephen, has said that WAK is a stage in the progress toward "a totally implantable artificial kidney." The WAK will shrink in size as new sorbents are discovered. When these sorbents are efficient enough and immune reaction is completely controlled, the implantable artificial kidney will appear. Within twenty years, Dr. Stephen expects that a potential kidney transplant recipient will be able to choose among human kidneys, animal kidneys, and implantable artificial kidneys in a variety of models.*

* The diversity of work in this field may be illustrated by reference to two other programs. In September 1980, the World Health Organization announced in Geneva that a miniature dialysis device, to be inserted under the skin, was expected to be available "in three years." In November 1980, *The Times* of London reported a form of treatment which involved the infusion of a solution into the peritoneum

Another group at the University of Utah Division of Artificial Organs has made the Utah Arm. Weighing 3 pounds (1.4 kg), this artificial limb is much lighter than a natural arm. It is nonetheless strong, silent, and efficient, reacting noiselessly and immediately to the myo-electrical signals given off by the patient's arm stump or shoulder muscles. The Utah Arm has been described as moving "as the patient would move his natural arm did he have one." One of its characteristics is an artificial muscle that is capable of contraction, thus shortening the arm. Its hand has metal claws. Utah is by no means alone. This type of development is also well-advanced in other parts of the United States, as at the Department of Bioengineering at Rancho Los Amigos Hospital in Downey, California, and at the faculties of medicine and science at Montpellier in France.

The Utah program has produced completely artificial hearts which have kept calves alive for as long as six months. Calves have been chosen because they are gentle animals, and because when very young, their hearts are about the same size as human hearts. Some of these artificial hearts are made of moulded polyurethane, others of moulded silastic and Dacron. Calves have been kept in the laboratory, but after removal of their natural hearts and the implanting of artificial hearts, they have been able to eat, drink, and exercise on a treadmill. One calf grew from 187 pounds (85 kg) to 297 pounds (135 kg) in four months with a polyurethane heart. The reason for its death was "a broken heart valve." Some of these hearts are powered by electricity and others by compressed air.

In more recent times, the Utah team has produced a feasible proposal for an atom-powered artificial heart. The concept involves implanting a pump mechanism inside the chest and powering it with a nuclear "thermal converter" in the abdominal wall. This project has been brought to a halt by bizarre problems not connected to medical or scientific difficulties: American laws concerning nuclear power, and

----

(the double membrane covering the abdominal cavity). A thin plastic tube is permanently fixed in the patient's abdomen and the solution is infused from a plastic bag connected to the tube. The solution attracts and holds blood impurities and four to six hours later is emptied back into the bag. The process is then repeated. This method, called "continuous ambulatory peritoneal dialysis," had been used for some four years in the United States and two years in Britain.

worry over kidnapping, terrorism, and blackmail of a person with an atomic device in his body have stopped further work.

Professor Kolff is optimistic on the long-term prospects of the artificial heart for human beings. He has more than once drawn attention to the American death rate from heart disease, now over 700,000 a year. Even with a major breakthrough in immunology and the removal of other current inhibitions upon the growth of human heart transplants, he considers that there will never be enough human hearts available to supply the demand. A human heart should be beating at the time of removal in order to confer the greatest benefit upon the recipient, and this imposes limitations upon the kind of donor who can be used for a heart transplant. For reasons mentioned earlier, motor-vehicle accidents are a prime source of donors, and fatalities in the United States now exceed 50,000 per year. Professor Kolff has suggested that the gap between availability and demand can only be filled by artificial hearts, which eventually will be "waiting on the shelf" until required.

The quest for a workable artificial heart is widespread. In Europe, Berlin has been described as "the capital of the artificial heart" because of the work of Professor Emil Bücherl at the Berlin Free University, dating from before 1960. Also using calves, Berlin has followed a path similar to that followed in Utah. In 1979, Professor Bücherl was reported as saying that his team has reached the stage where they could implant a polyurethane artificial heart in a man with an 80-percent chance of keeping him alive for up to two months. The purpose of the implant would be to maintain a patient awaiting a heart transplant whose diseased natural heart was likely to stop at any time. Because of the shortage of hearts at Stanford University Medical Center, where Dr. Norman Shumway practises, up to one-third of all candidates accepted for transplant die before one becomes available.

One of America's best-known medical lawsuits concerned an attempt in 1969 to do exactly what the Berlin group now proposes. The case followed the implanting of the first totally artificial heart into the chest of a human being. The surgeons were Dr. Denton Cooley of Houston, Texas, and his colleague, Dr. Domingo Liotta, who had built a mechanical heart. The patient was a man of forty-six with a history of heart trouble. In March 1969, his condition was deteriorating so rapidly that Dr. Cooley recommended a heart transplant, but no donor could be found. Dr. Cooley then suggested the temporary use of the artificial heart in order to sustain the patient. The patient agreed and the mechanical heart was implanted. It functioned for some sixty-four hours. Within that time, a natural heart became available and was transplanted, but it failed, and the patient died thirty-six hours later. His widow sued, claiming that the procedure which had been followed was no more than improper experimentation, that there had been an absence of informed consent by her husband, and that the two doctors were guilty of negligence. The trial took nine days in a federal district court, and at the end of it the judge directed a verdict in favour of the doctors. The widow took the case to the U.S. Court of Appeals, where she lost again on the unanimous judgement of a three-man bench.

Utah also has an artificial eye project, the aim of which is to enable totally blind people, including those without eyes, to "see." This well-advanced project experiments with blind volunteers, who already have "seen" light and certain patterns. The artificial eye functions basically by stimulating the visual centers of the brain (optical cortex). A person who has sight recognizes light as a result of messages or sensations transmitted from the retina of the eye through the optic nerve to the optical cortex. It has been known for some time that direct electrical stimulation of the visual centers of the brain causes a blind person to perceive spots of light. These points of light, like white dots or stars on a black background, seem to float in space at about arm's length. The artificial eye involves the screwing of a pyrolitic plug into the patient's skull, which acts like an electric wall socket. Through the plug, an "array" of up to 250 platinum electrodes, or tiny wires, is conducted in an insulated cable of Teflon so that they make contact with the "visual center" of the brain.

Outside the skull the electrodes are connected to a computer system that contains a camera. The camera transmits "pictures" to the computer, which searches its memory for recognition of the object, then transmits the shape of the object by means of electric current through the electrodes to the person. The person will then "see" the shape in the "white dot" fashion described above. The amount of his "vision" will depend on the amount of programming that can be fed into the computer. Vision of this kind cannot reproduce normal sight but will, it is hoped, allow a blind person to "read" written materials, to perceive shapes such as faces, and to distinguish features such as ears, eyes, and hair.

The Utah team hopes to produce a miniaturized device that could be entirely "worn" by the patient. The plug and electrodes would be permanently in place in the skull. The camera would be placed in a glass eye that sits in the eye socket and is attached to the eye muscles. The computer would be contained in the earpieces and frames of a pair of spectacles, and these would be connected to both the electrodes and the "eye camera" by wires. This "miniaturization" may be some years away but the device and the principles are already proven. It is now a matter of technology. Blind patients have already "seen."

Parallel work aimed at enabling totally deaf people to achieve some hearing is being conducted at Utah as well as at other institutions in the United States, France, and elsewhere. Provided that the patient's acoustic nerve is intact, the use of similar principles will result in the ability to hear certain sounds.

Astonishing things have already been produced and will continue to be produced in the field of artificial organs. Some of them, as mentioned earlier, have a science fiction quality. The thought of a human being with a nuclear-powered heart is startling, despite the television achievements of "The Bionic Man" and "The Bionic Woman."

As for its relationship with human tissue transplants, there seems to be no reason to look upon the field of prosthetics as anything but complementary. The same applies to the use of animal tissues. Despite suggestions by certain medical practitioners and writers that there is some kind of contest that will produce a winner, it is obvious that each has its place. One would imagine that questions of cost would inhibit the development and use of many prosthetics, and restrict general availability of the more complex ones. In 1979, Professor Bücherl estimated the cost of an "ideal" artificial heart with a suitable energy source at approximately 100,000 marks. Obviously, mass production would greatly reduce this sum. In the same year, the cost of a standard pacemaker in the United States was reported as $2,500. Only the future will provide definite answers, but one thing is very clear. In today's overpopulated world, there is no shortage of human bodies. The new question is not so much whether they are to be used as sources of spare parts, but how.

III.

The biggest single cause of failure in transplanted organs is rejection. "To reject" is defined by the dictionary as "to refuse to accept" and "to throw away." In the context of transplantation, the word has one meaning only, and every transplant surgeon knows precisely what it describes. Rejection refers to the fact that the body of each human being has its own means of recognizing the tissues of another human being and will reject such "foreign" tissue if it is implanted. There are some apparent exceptions to this general rule. If the tissues are essentially identical, rejection will not occur (this is the case with identical twins and with certain strains of highly inbred animals). Transplantation or grafting of a person's own tissues from one part of his body to another avoids the problem, too. There are also extraordinary features applicable to fetuses and pregnant women. Again, a few human body parts seem hardly to provoke the rejection mechanism at all, some, such as corneas and cartilage, because of their natural characteristics—the cornea is one of the few tissues that has virtually no stimulator cells, so the new cornea provokes little immune response from the host body and requires minimal support from immunosuppressive drugs when transplanted—and others because they are quite inert and sterilized before transplant, such as the ossicles, a trio of small connected bones of the middle ear which are transplanted to cure forms of deafness. Nonetheless, the general rule holds good with most body parts.

The body's mechanism of rejection is commonly called the immune system. Expressions such as "immune reaction," "immune response," and "immune rejection" all refer to this mechanism. "Immunology" describes the science or study of the phenomenon and is much concerned with "immunosuppression." In 1977, one of the world's most respected surgeons addressed a medical conference in Japan on progress in immunology: "Further . . . advance can be anticipated, and when it happens there should be a spectacular increase both in the volume and range of transplanted organs." The speaker had observed that in the five years before 1977 something of a plateau had been reached in clinical aspects of transplantation, mostly because there had been no major developments in the field of immunosuppression, and no substantial breakthrough to further reduce "the rejection of transplanted tissue." This is now changing. At about the same time, a leading immunologist wrote, with some understatement, "Clearly the major technical barrier to extensive transplantation is the problem of rejection. If this problem can be overcome we could expect a dramatic rise in what might be called 'spare parts' surgery. There may be a considerable problem associated with the availability of suitable tissues for transplantation."

It is well-known that every cell in the body carries material that can engage in the "immune response" struggle. It may be useful to try to describe briefly what happens when this contest occurs after transplantation of a body part. As is usual in warfare, there are two sides. One comprises the "foreign" or invading tissues, and the other the tissues of the recipient or "host" body. The first critical skirmish involves recognition by the recipient that the invaders are foreign. Like all human tissues, the invaders have unmistakable distinguishing features or cells (antigens and stimulator material) on their surfaces. These distinguishing features are promptly recognized as foreign by specialist cells (lymphocytes) in the host body. Each lymphocyte carries on its surface a kind of radar ("receptor molecules"), the function of which is to detect foreign stimulator-plus-antigens cells. Lymphocytes behave like a permanent military patrol circulating in the bloodstream. Once the stimulator material is recognized as "foreign," the host body will automatically attack all the transplanted tissue and bring into play a process that will result in rejection of the foreign tissue.

If there were no such phenomenon as immune rejection, hearts, lungs, kidneys, livers, skin, bone marrow, bones, intestines, pancreases, and a number of glands could be transplanted with the expectation of complete success. Because of immune rejection, the transplantation of every one of those tissues will fail, usually within two or three weeks, unless there is successful immunological control (or unless there is a perfect tissue match). The present state of medical knowledge is such that successful control cannot be expected in all cases.

The success of immunological control in transplants varies from one kind of tissue or organ to another. Today a kidney transplant has something like a two-in-three chance of being a long-term success.

With the heart the prospects are a little better. But the possibility of success with a cornea is much higher because immune rejection is not so important a factor.

Mathematically, a two-in-three chance of success may be high, but it still means that one out of every three transplanted kidneys and hearts will fail. For a person with two kidneys, the failure of a kidney transplant is much less inhibiting than for a person with one or none. Even a patient who has had both kidneys removed will usually be able to resort to the artificial kidney machine if he loses a transplant by immune rejection. Rejection is therefore not normally the catastrophe for kidney patients that it is for heart patients. A two-thirds success rate (even 70 percent) has a different aspect when offered to a heart patient as opposed to a kidney patient. If a heart transplant fails, the patient dies. Dr. Barnard's use of "piggy-back" heart transplants is designed to alleviate their all-or-nothing risk but still the risk is not comparable to that for kidney transplants because of the deteriorating nature of the original heart and the absence of any backup apparatus for hearts such as the artificial kidney machine.

Liver transplants have had a fairly high degree of success—by the late 1970s, it was only slightly less than the rate for kidneys—but the number of livers transplanted has been only about 1 percent of the number for kidneys. The liver is a vital, unpaired organ, but immune rejection is not so great a problem as it is with other organs and accounts for less than 10 percent of failures. Still, elimination of rejection problems would confer a double benefit on recipients because it would also eliminate the need to use immunosuppressive drugs, which reduce the body's resistance to disease and thus increase the number of posttransplant deaths due to infection. In October 1977, a team of researchers at Virginia Medical College announced a major improvement in the development of drugs used to fight rejection of transplanted livers and other organs. A year later, Dr. Thomas E. Starzl of Denver, Colorado, who has carried out most of the world's liver transplants (totalling over three hundred to that time), announced a new treatment that reduced the needed quantities of immunosuppressive drugs in liver, heart, pancreas, and lung transplants. "Don't call it a breakthrough," said Dr. Starzl, "because that gets people's hopes too high, but I think it is very significant."

By the end of 1979, further advances had been made with liver, pancreas, and kidney transplants. In England, as mentioned previously, the first two pancreas transplants were carried out by Professor Roy Calne at Cambridge. Professor Calne, who has transplanted more than five hundred kidneys as well as many livers, is regarded with Dr. Starzl as one of the world's leading liver transplant surgeons. Each pancreas transplant was combined with the simultaneous transplant of another organ. Professor Calne used a new rejection-control drug called Cyclosporin A, combined with surgical techniques developed in Lyons, France. The first operation involved transplantation of a pancreas and kidney, and the second a pancreas and liver. At the same time, it was announced by the Australian National University that a fifteen-year program of research in immunology had resulted in a significant breakthrough in the cure of diabetes in animals, using pretreated, transplanted pancreatic tissues, which caused no immune rejection. The successful use of this transplant technique with human diabetics is confidently expected.

The history of lung transplants is one of almost total failure, in large part because of immune rejection. The lung, however, is one of a pair, so the risks of surgery are not so dire to the recipient.

For all these reasons, patients have very different attitudes to transplants. One kidney patient may prefer to delay receiving a transplant until he can get a kidney with a high degree of tissue compatibility, while another, who is sicker or has an unusual tissue type, may be more anxious to proceed and more prepared to take risks. For neither of them will rejection necessarily be fatal.

On the other hand, a patient with heart disease and with only a few weeks to live does not have a choice in any real sense of the word. He may grab at any chance of survival. To him, a two-thirds prospect of success may seem extremely favourable odds. The cornea recipient will have a different attitude altogether, because the chances of rejection are virtually nil. The elements of persuasion and dissuasion for him will obviously be of another kind.

Still, the truth remains that rejection is a danger with the transplantation of most tissues. Elimination of the rejection problem would obviously reduce this danger or even banish it. If rejection were as unlikely for a transplanted heart, kidney, gland, liver, lung, or pancreas as it is for corneas and cartilage, then it would seem that the eradication of many presently incurable diseases would be prevented only by shortages of human body parts, skilled medical personnel, and hospital facilities.

Even under present conditions, the community need for human tissues enlarges ceaselessly. It is impossible to quantify the increase in that appetite that will occur when immune response ceases to be an impediment to therapy. There is more to it than the mere exclusion of failure in a given proportion of transplant operations. Confidence in transplants of vital organs could lead to a huge jump in demand. It requires little imagination to visualize the likely difference in attitude among heart disease patients if, instead of failing once in three cases, heart transplants could be expected to succeed in every case.

In the United States, an effort was made in 1969 to quantify the demand for hearts that would follow the elimination of immune rejection. A National Health Institutes Task Force concluded that once this "major roadblock" was removed, the number of heart transplants would rise from an assumed base figure of one hundred transplants in a year to about twelve thousand, with a further estimated increase to thirty-two thousand if a satisfactory circulatory assist device and an artificial heart were also available.

What can be done to control the immune response? What directions are being followed in research?

Scientists have indicated three broad avenues. The first attempts to find donors and recipients with similar or compatible tissues, in the same way as is routine for blood transfusions. The process is called "tissue typing" and has given rise to expressions such as "tissue compatibility" and "histocompatibility." Unfortunately, tissue typing is a far more complex matter than blood typing. The prospects of finding a person whose blood group is compatible enough with the recipient's to permit acceptable transfusion is one in six or better. The prospects of finding a person whose tissue is sufficiently compatible with the recipient's to predict, say, a successful kidney transplant is about one in sixty thousand. Apart from this, the technology of tissue matching is still not perfected. Even so, it is sufficiently sophisticated to get fairly good results with transplants (of kidneys, for example) when tissue typing is used in conjunction with drugs designed to suppress the function of the recipient's immune system.

The use of these immunosuppressive drugs is the second method adopted by scientists in tackling tissue rejection. The drawback is that these drugs can have a drastic effect upon a patient because they so lower his resistance to disease that he may be suddenly overwhelmed by random infection. There are other dangers, too. When Cyclosporin A was first used in Cambridge in 1978, it was hailed by some as possibly the most important advance made in controlling immune rejection without drastic effects upon patients; but after it had been in use for some fifteen months, Professor Calne and his colleagues published an article in *The Lancet* calling for caution before giving it general usage because a form of cancer had occurred in the lymphatic systems of three of the thirty-four patients who had been treated with it to that time.

The third avenue of research is more recent. One side of it involves treating the tissue to be transplanted so that it will not activate the recipient's immune system. In experimental work with animals such as mice and rats, the stimulator cells have been suppressed or removed from tissues prior to transplantation; they have therefore not provoked the immune response, and the tissues have functioned for long periods of time instead of being promptly rejected. In recent Australian work with diabetes, the treated tissue has been accepted by the host body and has successfully reversed, i.e. cured, the diabetes by producing insulin as needed. The advantage of this approach is that it does not require giving the recipient massive debilitating doses of drugs.

The other side of this research devotes attention solely to the host body and aims at suppressing the reaction of the lymphocytes that respond to specific types of invading tissue. The host body's cells are treated in order to neutralize the "receptor molecules," so that the body's ability to recognize invading or foreign tissue is inhibited. This has succeeded experimentally in Sweden in rats.

All of this research will lead to continued improvement in organ transplants and the long-term survival rates of recipients, and will open up new areas of transplantation—for example, of glands such as the parathyroid and the pituitary, as well as the islet cells of the pancreas (which produce insulin needed by diabetics).

In many parts of the world, immunologists have devoted close attention to the human fetus. Fetal thymus cells and bone marrow are particularly important in research on certain diseases of infants in which the normal mechanism for resistance against infection is deficient. Fetal cells are also used in research into rejection of liver and kidney transplants in adults, and for tissue typing in transplant surgery. Fetal thymus glands are frequently transplanted to children. Some medical practitioners have expressed the view that fetal body parts, when transplanted, do not cause so severe an immune reaction as normal tissues, although this opinion has been rejected by leading immunologists.

The University of Aberdeen in Scotland has conducted research into the relationship between pregnancy and transplant rejection. A specialist in reproductive endocrinology from that university suggested in a public statement in November 1977 that the study of pregnant women may lead to significant advances in the control of rejection. The question is: Why, since half of the genetic material comprising the embryo or fetus is contributed by the father, doesn't a pregnant woman's body reject what is genetically foreign tissue? So far from rapidly rejecting it, the mother's body does the opposite, and firmly holds an embryo. The doctors of Aberdeen suggested that the large quantities of protein-related substances produced by the placenta and poured into a pregnant woman's bloodstream may be the cause of the immunosuppressive response which allows retention of the fetus. The possibilities opened up by this line of research range from the future development of a "simple vaccination" against pregnancy (a kind of contraception or abortifacient) to an immunosuppressive agent to control rejection of transplanted organs.

There seems to be every reason to believe that current immunological research will meet with success. Fruitful long-term programs are being carried out around the world by research teams as far apart as Uppsala in Sweden and Canberra in Australia. Improvements are constantly appearing. As substantial breakthroughs occur, the source of the materials that will be called for in much greater quantities will be the human body. One American expert, a director of surgical research at Harvard Medical School and Peter Bent Brigham Hospital in Boston, threw caution to the winds a few years ago at an international symposium on the future of transplantation with these words: "The practical possibilities are limitless; progress is so rapid, the need so great, and the interest so high that it would be rash to put any limits to future progress in organ transplantation."

IV.

Many people prefer to put their heads in the sand when it becomes necessary to discuss money in the provision of medical services. "How can you be so callous as to weigh life and health against money?" is a common sentiment. Politicians and professional people often try to avoid public debate on this emotive topic. Yet to carry a process from initial research and experiment through to practical application can require huge sums of money.

The first heart transplants were expensive almost beyond calculation because they effectively brought entire hospitals to a halt and demanded the availability of virtually all the staff and facilities. By 1980, the English Department of Health was able to estimate the cost of a heart transplant at £17,300 for the surgery and the first year of follow-up care. Livers were considered to be at least as costly as hearts. One leading English heart transplant hospital claimed that its transplants were costing no more than £13,000 for the first year and £2,000 for each subsequent year. Other official English estimates were: cornea transplant including ten days in the hospital, £750 (1979 costs); kidney transplant, £4,000 for surgery and first year and £500 for each subsequent year (1978 costs). By way of contrast, the English cost of hospital dialysis for kidneys was much higher than transplantation, standing at £12,500 a year in 1977 and decreasing in the following years as the efficiency of the machines improved. Home dialysis was estimated at £8,000 a year, with similar decreases in subsequent years. In 1978, the total number of registered British kidney patients on dialysis was about three thousand, giving an overall maximum national cost for that year of about £30–40 million. As for the United States, estimates from the medical faculty of a well-known university in 1979 put the average cost of all dialysis at roughly the same annual amount as England, $24,000. With fifty thousand patients under treatment, the national figure was $1.2 billion. Other annual American costs were up to $50 million for the use of heart-lung machines based on one hundred thousand surgical operations, $125 million for fifty thousand heart pacemakers ($2,500 each) excluding surgery, and for heart operations not known or used until 1970, $1 billion. In a society that has socialized medicine, such as Britain, the patient pays no direct fee for surgery or hospital treatment; the cost is borne by the community. The liver transplant recipient in England will pay nothing, while in the United States he may have to find more than $50,000 for the surgery alone. In September 1979, an English couple began a public appeal to enable them to travel to America and pay Dr. Starzl, in Denver, for a liver transplant to their small son. The surgeon's fee was reported as exceeding $55,000. After the appeal was launched, Dr. Starzl announced that he had decided to waive his fee and perform the operation free of charge.

It is plain that the ability to obtain unusual and expensive medical treatment is governed by monetary considerations under socialized medicine just as it is under "free enterprise" medicine. The differences are in the identity of those who must bear the high costs. When a government that presides over a socialized medicine program announces public spending cuts, as the British government did in 1979, the effect is soon noticed. A South London kidney unit with sixteen waiting recipients announced in November of that year that it had suspended its transplant program. Transplants were resumed only when a Kidney Patients Association advanced £5,000 per patient to the unit. In January 1980, a public row broke out over the use of a particular immunosuppressive drug, Pressimmune, at the same hospital. The first-year cost of this drug per transplant patient was £6,000, increasing the overall cost of a kidney transplant to £10,000 as compared to £5,000 when other drugs were used. Organ transplant units all over Britain found that the shortage of government grants was forcing a reduction of their programs. It was revealed in February 1980 that new models of heart pacemakers capable of giving patients a near-normal heart function would not be supplied in Britain because no public money was available to make or acquire them.

Statistics graphically illustrate the power of governments under socialized medicine to control medical services. The numbers of European patients allowed to receive treatment for kidney disease of the kind that calls for dialysis and transplant vary markedly from nation to nation. In Britain, the number of patients alive and receiving such treatment at the end of 1978 was 94 per million, while in Switzerland it was 198. Of fifteen nations registered with the European Dialysis and Transplant Association, Britain was twelfth in its provision of these facilities. One of the clearest demonstrations of government control is offered by Holland, where both dialysis and kidney transplants are forbidden without a license from the public health authorities. Regulations prescribe the volume of services—for example, in 1974 a maximum of 71 dialysis facilities and 32 transplants per annum per million people were permitted. By the end of 1978, the combined figure had risen to 130. Another form of control, which is practised in England, is the deliberate exclusion of patients because of their age: people under the age of fifteen and over fifty-five are not admitted to such kidney treatment. This horrifies some critics. The president of one of England's kidney patients' associations complained publicly in February 1980 that this policy, which has no valid medical basis, was allowing the deaths of over three thousand untreated citizens a year.

Nothing is more potent than the power of the purse when it comes to the provision of medical services, especially the innovative and expensive ones. The richer the community, the more opportunity it has to acquire the best and the latest. Poor nations and those lacking free, well-financed research are likely to be unable to make the best use of expensive new medical knowledge whether it concerns the utility of human tissues or anything else. Economic considerations control the quality of all services available in all societies, and medicine is no exception. Therefore, just as transplants and other medical procedures that use human tissues for therapeutic purposes can be reduced or even eliminated by lessening the money supply, so can other services. In assessing future community claims upon the body, economics is a vital consideration, but it is of a different nature entirely from the other influences.

CHAPTER 3

The History of Legal Regulation

Whether alive or dead, every body stands today as a potential source of therapeutic benefit to the body of every other human. What sort of procedures are needed to ensure that the sick secure this benefit? Are public laws the most practical way of laying down procedures? What laws do we have already and what have they accomplished? Do we need laws at all?

A reading of transplant laws in Western countries discloses that they are primarily directed not so much toward surgical operations or medical techniques as toward the materials used—human tissues. It is the taking from bodies of some of their contents that disturbs and attracts the concern of society and lawmakers. But one must ask whether it is really necessary to create laws solely to regulate the activities associated with the removal and therapeutic use of human tissues. Perhaps the community and its medical profession could achieve a good working relationship without a surrounding web of legal regulation. If a new medical procedure produces reasonable results in treating sickness and attracts public patronage, why should the law step in? Many responsible doctors strongly resent official restriction.

Surprisingly, there is substantial medical precedent for a claim of noninterference by government legal rules. The precedent is provided by modern surgery itself, which commenced with the development of anaesthetics and antiseptics in the mid-nineteenth century. In the West, particularly in the English-speaking democracies, we have a remarkable dearth of written law regulating the performance of surgery.

Of course, precedent alone should not be decisive, even in matters of law. If it were, surgeons would be put to flight by the most ancient legal precedent of all, a code fashioned by Hammurabi, the Babylonian king who ruled somewhere between seventeen and twenty centuries before the birth of Christ. One of the most famous codes of law known to man, and certainly the oldest, the Code of Hammurabi contained nine paragraphs (215–223) that closely regulated surgery. Two of them provided that a surgeon who used his "lancet of bronze" and saved the life or the eye of a free man became entitled to ten shekels of silver (but only two shekels if the patient was a slave). With impeccable judicial logic, two later paragraphs provided that if the surgeon caused the free man's death or destroyed his eye, "they shall cut off his forehand" (if he caused the death of a slave his obligation was to "replace slave for slave," and if he destroyed the eye of a slave he had to pay half the slave's price in silver).

One perceptive writer has conjectured that this law might hold little fear for a modern surgeon. Suffering the temporary inconvenience of amputated hands, he would no doubt ask those of his colleagues who specialized in transplants to stand by and be ready to graft to him the hands taken from the next medical offender who attracted the justice of Hammurabi (provided they were fresh enough). The newly severed parts would not be wasted, and the best would be made of a bad lot.

Many people have been puzzled by the tendency of the elected governments of nations such as Britain and the United States to rely, as the source of law on surgery, on a few ancient principles laid down by judges in bygone days on subjects unrelated to the practice of medicine. There is a great deal to be said for the promulgation of guidelines designed to cover the more worrying dilemmas of the operating theatre. Surely the law should be in a position to offer straight answers to both surgeon and patient—or is it preferable to leave the resolution of disputes and questions to a court applying and adapting broad principles? Should a surgeon be liable to punishment under the criminal law or to pay damages under the civil law (or should he be liable to both) in the following cases?

- He opens a patient's skull and removes a brain tumour with the patient's prior knowledge, consent, and understanding of the risks. The operation is not successful, and the patient is permanently disfigured. The surgery is performed competently.

- During the delivery of a child by Caesarean section, he discovers tumours in the walls of the patient's uterus. Worried about the consequences for her if she became pregnant again, he ties off her fallopian tubes. This is done solely for the patient's benefit, but without her knowledge or consent. The surgery is performed competently.

- He performs major emergency lifesaving surgery upon an unconscious accident victim. He amputates a leg, and removes an eye and a kidney. The patient first learns of the surgery when he wakes up. The surgery is performed competently.

- He performs corrective eye surgery upon a ten-year-old child without the consent of the child but with the consent of the parents. The operation fails and the child is permanently disfigured. The surgery is performed competently. Would it have made any difference if the patient had been five, or fifteen, or had been a mental incompetent?

- He performs corrective eye surgery upon a ten-year-old child without the consent of the child or the parents. The surgeon knows well that the parents strongly oppose surgery on religious grounds. The operation is a complete success and the child is saved from a lifetime of defective eyesight, but the child's life was not at risk. The surgery is performed competently.

- He performs a multilating operation with the prior knowledge and consent of the patient—in fact, at the patient's pressing request. The operation is to amputate the patient's forearm because he has a chronic skin rash on it, and an overpowering desire to make sure that he will never be conscripted for military service. The surgery is performed competently.

The surgeon is obviously vulnerable legally in some of these cases, and the number of possible comparable cases is virtually infinite, just as the number of surgical operations performed in the past century and a half is virtually incalculable. However, the startling fact remains that in the advanced countries of North America and the British Commonwealth, there exists no generally applicable law or code that provides clear answers.

Laws aplenty regulate medical negligence, as American doctors know only too well. The law of negligence, however, deals with the *quality* of surgery. It is concerned with whether the doctor failed to act with the care and skill that is to be expected of him. Accordingly, the law of negligence will penalize any person who owes a legally recognized duty to another and fails to perform it competently, be he butcher, baker, candlestickmaker, or doctor. The question, though, is not whether there is a law controlling negligent surgery (there is!), but whether there is a law allowing surgery to take place at all, and if so, on what conditions. If there is no such law, should there be one? Assuming a surgeon to be competent and concerned to help the sick and injured, should he be more or less free to wield his knife whenever he decides it is proper to do so—for example, against the wishes of the parent of our ten-year-old patient, or with their permission without reference to the child? As Lord Devlin, the famous English judge, has put it, the direct law of surgery "suffers from under-development," while the law of medical and surgical negligence "suffers from over-development." Contrasts could be made with many other professional and business activities, but one will serve at present. An airplane pilot must not only be highly qualified and legally licensed before he may take an aircraft off the ground, but must obey many other laws and directions before he may do so on a particular flight. In addition, he will be subject to legal controls (but at the same time will know the extent of his powers and privileges) while in the air, dealing with passengers, and performing other functions. Overriding all of this will be the law of negligence, for the pilot must discharge all his duties with proper competence if he is to avoid further liability.

When hypothetical cases are described and leading questions are asked, as they have been above, many people quickly answer, Yes, of course there should be control over doctors; of course there should be a code of laws. However, for the past 150 years no new set of legal principles has been created in the West to lay down ground rules for modern surgery. In the English-speaking countries, there has not only been no public demand for such laws but virtually no litigation either. In North America, Britain, and Australia, one is hard put to find any lawsuits that discuss the limits of a surgeon's legal rights and duties. In both the United States and Britain, the closest pertinent lawsuits have involved extraordinary if not grotesque behaviour, such as sexual aberration and nineteenth-century prizefighting, rather than surgery. One of the rare English lawsuits of modern times that did involve a medical process was a divorce suit in 1954 in which a husband and his doctor were strongly criticized by one of England's most eminent judges, Lord Justice Denning, for their respective parts in carrying out the husband's sterilization at the husband's request. The judge was firmly of the view that the common law of England prohibited such an operation, despite the husband's consent and wish for it, and made the surgeon liable to a criminal charge of assault and battery unless there was "just cause or excuse" such as preventing the transmission of an hereditary disease. The first leading English case was one in 1882, in which a court of four judges decided that the combatants in a bare-knuckle prizefight were guilty of criminal assault and battery upon each other despite the obvious fact that they each agreed to take part in the fight. In 1934, the English Court of Criminal Appeal upheld the conviction for indecent and common assault and battery of a man who obtained sexual gratification from thrashing a girl with a cane with her willing consent. The court held that the man's intention had been to do bodily harm and therefore the girl's consent was ineffective. The leading American case, much quoted since it was decided in 1948, involved an army lieutenant who burned his initials with a lighted cigarette on his girl friend's breasts, thighs, and buttocks. She also consented, but the court ignored her consent and convicted the soldier of assault with intention to maim.

One might justifiably ask what such *outré* circumstances and decisions can possibly have to do with the conduct of a medical practice and the limits of lawful surgery. It is certainly difficult to draw any persuasive parallel between these cases and the hypothetical ones described earlier. Yet in the absence of articulate rules prescribed by legislatures or by departmental regulations, the best that the best legal brains can do is to suggest that the law of modern surgery is based on mediaeval principles as adapted by such cases as these. Addressing an international symposium on law, medicine, and transplantation on the eve of the first successful heart transplants, an eminent lawyer put it inimitably:

> In most systems nowadays an ordinary operation is still looked on as being in the nature of an asssault and battery, *Körperverletzung* [bodily injury], or

something of that sort, justified or legalized by consent or its equivalent. . . . [This] is certainly by now unrealistic. An operation should be treated as a positive, beneficent, admirable action from the outset, not as a lawful infliction of harm. . . . After all we do not construe marital, or nowadays even extramarital, intercourse as rape licensed by virtue of consent and perhaps registration. Or if my cook makes zabaglione with my eggs, this is not commonly thought of as destruction of the material rendered lawful by his office.

Some American and English lawyers claim that modern variants of the medieval crime of "mayhem" (from which the word "maim" is derived) are also applicable to surgery. Mayhem or maim is the act of permanently disabling or weakening a man, "rendering him less able in fighting." The leading case was decided in England in 1604 when (according to the law reports) a "young, strong and lustie rogue, to make himself impotent," arranged for his left hand to be cut off by another man so that he could make his living as a beggar. They were both convicted. The source of the crime was the duty that men had in feudal times to fight for their lords and, ultimately, the king. Because of this, nobody had the right to maim himself voluntarily. Similar reasoning made attempted suicide a crime for many centuries. By analogy, a surgeon who cuts open a patient will commit the offence of maim, or assault and battery (maim is also known in the United States, which acquired the concept from England and developed it), unless he can show not only that the patient knowingly consented to the procedure but also that it was for his benefit. The common law has never permitted a citizen either to inflict serious injury upon himself or to agree to some other person doing so, unless there is an accepted excuse or a countervailing benefit.

In the United States, the laws on surgery are further complicated by variations among states. Most American jurists adopt the assault and battery principles as a starting point, and accept the idea that a person cannot lawfully consent to being killed by another, or to having serious bodily injury inflicted upon him. However, the limits of a person's power to consent to bodily injury to himself are not very clear. A model penal code prepared by the American Law Institute in the 1960s supplied rational guidelines to regulate surgery, but this code is an ideal objective rather than a reality.

Unfortunately, any serious attempt to mould these ancient princi-

ples to modern surgical conditions runs quickly into trouble, as would
a blacksmith who began to make metal parts for jet aircraft. One
source of trouble is the law governing sport. Legal principle has always
forbidden a man to agree to allow another to injure him, and punishes
them both if injury is inflicted. Yet the law has been obliged to back-
pedal with body-contact sports such as boxing, wrestling, and football,
where playing the game is not necessarily beneficial to the player and
does necessarily involve the risk of physical injury. The legal limits of
"consent" to injury in sport have never been satisfactorily defined
either, and are only in slightly better condition than their counterparts
in surgery.

In relation to surgery, the borrowed principle that requires the patient's consent in addition to a benefit for him cannot be applied to the
emergency case where he is unconscious, or to an emergency that develops during an operation. Again, the requirement for "consent" becomes difficult to obey in the case of children. To what extent can you
say that a consent is truly given to surgery upon a child when it is the
consent of the parents and not that of the child? At what age does one
draw the line and decide that the child's own consent may be given
while its parents' is not needed? The questions could be endless, and a
comprehensive code would be needed to give all the answers—yet they
rarely seem to arise in real life as far as the law is concerned. In both
surgery and sport, it is, for lawyers, rather embarrassingly apparent
that the community has moved comfortably along without specially
designed legal rules.

One might be excused for imagining that the courts would have
been choked with cases seeking to clarify and define the precise extent
to which a doctor may make incisions with his scalpel, and a sports-
man may inflict physical damage upon an opponent. There have been
hardly any. When surgeons are told by judges that the exercise of the
latest surgical techniques could result in criminal liability under long-
dormant laws created to protect seigneurial rights in the Middle Ages,
such as the law of mayhem, it is not surprising that they listen politely,
shrug their shoulders, and go back unconcernedly to their operating
theatres. They know that the community feels the same way. A perceptive comment upon this state of affairs is that it expresses the public's genuine confidence in its medical personnel. It has been suggested

that the reason why there are virtually no specific modern laws on surgery is that none have been needed; despite the horrifying misdeeds of surgeons which the imagination can conjure up, the medical profession has behaved with responsibility in using the knife; the social worth of surgery does not need special proof. Had the community been dissatisfied with its surgeons (as opposed to their deviations from a high standard of competence, which are governed by the principles of negligence), there would have been more litigation, and the judges would then have been given the opportunity to create law. The same kind of comment has been made about sport; that sports and sportsmen, by and large, seem to be able to manage their affairs perfectly well.

On the other hand, those who do not hold medical practitioners in such high esteem may claim that firm and comprehensive laws are very much needed. A strong body of opinion holds that much unnecessary surgery continues to take place. In recent years, there has been widespread adverse comment on the massive incidence of radical mastectomies and hysterectomies, for example, in a number of Western nations.*

This glance at the laws of modern surgery has shown that in at least one substantial medical area, society has not insisted on a separate legal code. The question now is whether transplantation is another such area, and whether the removal of human tissue from the living and the dead should be left to be worked out between society and its doctors. Perhaps as a practical matter the question has already been answered: widespread legal regulation of transplantation is an accomplished fact, and the laws are likely to proliferate. Even so, there are other, more persuasive answers. One is that the removal of body parts from the dead stirs deep emotions in us all and is too disturbing to be left "to be worked out." Rules are needed from the outset, to protect everybody concerned, not least the medical practitioner.

Other reasons why specific laws on transplants and usage of human tissues might help rather than hinder have to do with the inadequacies

---

* Judging from the legal summary in the Council of Europe Report of 1975 there seems to be no designed code of law in Europe regulating surgical operations. The report indicated that the subject is normally governed by "general principles of civil and criminal law" and medical ethics.

of the so-called laws of surgery in dealing with even the most straight-forward transplant questions. One cornerstone of this law is the proposition that surgery must be for the benefit of the patient. This has provided an effective answer to many critics, even in the emergency wards, where patients' consent is "deemed" to be given to the beneficial treatment. But it is difficult if not impossible to describe the removal of tissue from a living person for transplant to another as "benefiting" the person from whom the tissue is taken. Direct law would eliminate confusion and authorize reasonable transplant procedures, because the overall benefit to the sick outweighs the detriment to the donor.

Again, the rapidity with which medical and scientific knowledge is growing suggests that the day has passed when society can afford to wait for lawsuits to appear so that judges may be given the opportunity to make new principles. Transplantation and therapy by means of body materials raise issues too profound to be ignored or dealt with on a random basis. We are dealing here with new concepts of life and death, and with unprecedented usage of the human body. Unless clarification is made, and quickly, grave social damage could be done, and worse, unacceptable practices could quickly appear that might prove impossible or very difficult to eliminate. On a less profound level, a legal vacuum can work against everybody's interests. Because of the law's failure to speak plainly about transplantation, eminent medical practitioners who have participated in heart and organ transplants in Japan, Sweden, the United States, and England have been accused of homicide, sued in civil courts, and involuntarily embroiled in inquests and criminal cases because of claims that their surgery constituted an unlawful activity.

Around 1950, legislatures the world over began to show the first signs of activity on this front, producing the first laws that regulated donation and use of dead bodies and body parts for medical and scientific purposes. The initial momentum for this kind of law was provided by the wide international acceptance of the worth of cornea transplants.

The Americans and the French began to move earlier than most. On May 1, 1947, Earl Warren, then Governor of California, approved what may well have been two historic statutes. Bearing the unprepossessing titles "Chapter 125" and "Chapter 126," they wrote a new charter whereby a citizen of that state could dispose of his or her own body or separate parts of it on death. Californians were authorized to give "the whole or any part" of their remains to a hospital, university, or similar institution, and their families and representatives were legally obliged to obey such a direction. The instructions could be contained in a will or "other written instrument." There was no stated limitation on the purpose for which body parts could be given. The law was wide enough to permit gifts of body parts for transplantation and other ways of healing the sick.

These statutes were subsequently modernized by amendments in 1957 and 1968. The 1957 amendment introduced provisions enabling gifts of body parts to be made to any eye bank, artery bank, blood bank, "or other therapeutic service." The 1968 amendment created a more elaborate machinery for the donation of human tissues for therapeutic purposes, and was, in effect, a high-water mark of state legislation before the Uniform Anatomical Gift Act of 1968 was presented to the American public for national adoption. California repealed its own legislation and adopted the Uniform Act on November 23, 1970.

Although the Californian initiative began in 1947, the bulk of the "first-growth" American laws did not appear until the 1950s.

The French also produced an early "transplant" law in 1947. By a decree of October 20, doctors in a limited number of listed hospitals were authorized to perform autopsies and remove tissues from dead patients "in the interests of science or therapeutics . . . even in the absence of the family's authorization." Apparently this decree later came to be regarded as a source of authority for the removal of organs for transplantation because of the words "in the interest of therapeutics." The justification for such an interpretation seems doubtful for a number of reasons, the first being that it was a "decree" (not a public statute) and that it was not part of a law on "interment, exhumation, cremation and transport of corpses." Secondly, it strains credulity to accept that this decree deliberately introduced, in 1947, a law to allow removal and transplant of body parts (not just organs) "without the family's authorization," so long before the appearance of viable organ transplants. It appears in fact to have been concerned solely with facilitating official autopsies and allowing the medical profession to retain and use for community benefit any tissues removed during autopsy. In any case, the likelihood that it was designed to facilitate transplantation is negated by the restricted terms of a French statute that appeared less than two years later. This public law (No. 49.890 of July 7, 1949) authorized, in the words of its own title, "the practice of corneal grafting with the aid of voluntary eye donors." Notice the restriction to a voluntary gift and to transplantation of one cadaver body part, the cornea. This was consistent with Western medical developments and social attitudes of the period. Still, limited as it was, the 1949 Act was among the earliest of its kind.

The English passed their Corneal Grafting Act in 1952, and this was quickly followed by similar Canadian and Australian state and provincial laws. Many of these, in order to facilitate the gift of eyes for corneal transplantation, used the same procedural mechanism that has been created by the English Anatomy Act of 1832 for giving bodies to medical schools for anatomical examination. South Africa, and Syria also, enacted corneal statutes in 1952. None of these laws was directed to the removal of body materials from the living.

Italian laws on the removal of human body parts had a bizarre origin. In 1940, Italy inserted in its Civil Code an article specifically forbidding the removal from any *living* person of "any part or organ" of his body the loss of which would permanently diminish physical integrity. This was the result of a much-publicized incident in the 1930s when a rich Italian citizen bought, after some negotiation, one of the testicles of a poor young Neapolitan and had it transplanted to his own body by a surgeon. Reports of the event do not indicate the result of the operation. Such transactions in the interest of recapturing virility were not unknown at that time. William Butler Yeats, who died in 1939 at the age of seventy-four, is said to have received in his later years the transplant of a monkey's testicle and to have claimed to be greatly rejuvenated by the operation (G. Fraser, Yeats' biographer, asserts "that it increased the fine eroticism" in his subsequent verse). Despite the geriatric justification for such surgery, Mussolini's Italy was outraged by the tale of the young Neapolitan, and the government responded with a blanket prohibition and criminal law punishments. To this day, Italian lawmakers have remained statutorily obsessed with genitalia. In 1967, the Italian law was relaxed to permit living persons to donate kidneys, but only on strict conditions.

As for tissue which can be removed from corpses in Italy, a law of April 4, 1957, first prescribed rules under which eyes could be lawfully removed for cornea transplants. By successive presidential decrees in 1961, 1965, and 1970, the scope of this law was enlarged to permit removal of specified tissues and organs. A provision that has characterized all Italian laws on cadaver-tissue removal since 1957 is a prohibition of the removal of genitalia. A new law was introduced in December 1975 to regulate cadaver transplants. This complex legislation envisages that objection—by the deceased in his lifetime, or by his relatives within a few hours prior to death—will prohibit tissue removal. But there are circumstances in which body parts may be taken without consent—for example, from corpses subject to autopsy.

This early wave of laws in the 1950s represented the first coherent attempts to support and regulate modern advances in transplant surgery. Often the laws were tentative and had little prospect of achieving what should have been one of their prime purposes—increased organ donation. Most of them dealt with tissue removal from dead bodies only. In some, the mechanism for obtaining consent to the removal was quite impractical, lacking a formula to resolve disputes among surviving relatives. A good example is the English legislation, which was copied by a number of other countries; it has led to many conflicts of opinion and remains confused to this day.

The first modern English law was the Corneal Grafting Act of 1952. This was replaced by the Human Tissue Act of 1961, which is still in force despite frequent moves to secure the passage of new laws. Under these statutes it has been impossible to decide with certainty who has the power to authorize removal of body parts from a person who dies in a hospital without indicating his wishes, because they conferred the power upon "the person lawfully in possession of the body" without bothering to identify exactly who is meant. Actually, this venerable and mysterious "person" was adopted by the Corneal Grafting Act at 120 years' removal from the Anatomy Act of 1832, where he was called the "Party having lawful possession of the body of any deceased person." Who is entitled to lawful possession? The family, the executor of the will, or the hospital? Or is it someone else altogether? This and other ambiguities have produced a series of public disputes in Britain which have adversely affected organ supply. In 1961, the Department of Health and Social Security stated in a circular that when a patient died in the hospital, the board of governors of the hospital could be regarded as "the person lawfully in possession" of the body, with power to authorize organ removal. Fourteen years closer to 1984, the Department expanded its view (which is no more than an opinion, and not necessarily a correct statement of law) in a 1975 circular which accepted the "area health authority" as another "person" with the necessary power. The 1961 circular had been undermined in 1973 when the Minister for Health in Parliament cast doubt on the validity of both hospital powers and donor cards, which were then in use in England. In 1974, Professor Roy Calne announced that kidney transplants had ceased at Cambridge because of the lack of kidney donation. According to a comment in the *British Medical Journal,* this followed from the attitude adopted by the Cambridge coroner, who insisted that the consent of a relative must be given in all cases, even those in which the deceased himself had left a written request that his organs be used for transplantation. Widely publicized arguments of this kind not only failed to resolve all the legal ambiguities, but they succeeded as well in reducing the supply of human tissues for treating the sick and increasing the number of deaths in patients who would otherwise have received transplants.

Other early transplant laws displayed little understanding of the urgency of many transplant requirements, and emphasized donations by will or testamentary document rather than by simple written or verbal means. Anachronistic references to wills still appear in many statutes including the Uniform Act in the United States, although that Act does envisage other methods of giving tissue.

The rapid and inexorable advance of medical knowledge soon made the early laws out of date. The viability of kidney transplant surgery from 1954 onward provided the first basis for change. A decade or so later, the profound effect of heart transplant surgery made change absolutely certain. One of the most successful projects in the regulation of transplantation was carried out in the United States between 1965 and 1968 by the Commissioners on Uniform State Laws. The Commissioners, who have been functioning since 1892, fill a gap in the American federal system. They prepare model laws on topics that are outside federal lawmaking power but within state power, and recommend them for adoption by the states, the purpose being to obtain uniformity on important matters and eliminate the confusion and uncertainty that flow from having conflicting state laws on the same subject. They have produced many model statutes, ranging from an entire commercial-law code to regulations governing the treatment of alcoholism.

In 1965, the Commissioners on Uniform State Laws appointed a subcommittee to study and report on the question of gifts of human body materials. By that time, a majority of American states had enacted their own laws (we have already seen the California product). These laws varied in coverage and quality, creating uncertainty whether a body-tissue gift made in one state could have any effect in another state. There were substantial differences about the identity of persons who could authorize tissue removal and make gifts, permissible recipients, the purposes for which the donated parts could be used, the procedure for making a gift, the formalities for revoking gifts already made, and the extent of protection afforded doctors acting in good faith under a gift document. The subcommittee found "both the common law and the present statutory picture . . . one of confusion, diversity and inadequacy." The most unfortunate result of this confusion was a cessation of the growth of gifts of body parts.

By 1968, the Commissioners had produced some order out of chaos in the form of a model Uniform Anatomical Gift Act. This model law, approved in July 1968 at their National Conference, was promptly endorsed by the American Bar Association and the appropriate committee of the American Medical Association. So widespread was the interest in the Act and its subject matter that three states had adopted a tentative draft in 1968 before the final draft was approved by the National Conference. Within five years it had been adopted by all fifty states and the District of Columbia, and it is still the last statutory word in the United States on transplantation.

The principal provisions of the Uniform Anatomical Gift Act are straightforward:

1. Any individual of sound mind and eighteen years of age or more may give all or any part of his body . . . the gift to take effect upon death.

2. In the absence of a gift by the deceased, and of any objection by the deceased, his or her relatives, in a stated order of priority (spouse, adult children, parents, adult brothers and sisters, etc.) have the power to give the body or any of its contents.
3. The recipients of a gift are restricted to hospitals, doctors, medical and dental schools, universities, tissue banks, and a specified individual in need of treatment. The purposes are restricted to transplantation, therapy, research, education, and the advancement of medical or dental science.
4. A gift may be made by will (to be effective immediately upon death without waiting for probate), or by a card or other document. If the donor is too sick or incapable of signing, it can be signed for him if two witnesses are present. A gift made by a relative can be made by document, or by telegraph or a recorded telephone message or other recorded message.
5. A gift may be revoked at any time.
6. A donee may accept or reject a gift.

Because of the infinite variations in circumstances, the Act gives blanket protection against lawsuits and criminal proceedings to any person "who acts in good faith in accord with the terms of this Act."

Despite its immediate widespread acceptance in the United States, the Uniform Anatomical Gift Act has been criticized. Some have been disappointed because it deals only with dead bodies and makes no attempt to regulate gifts of body materials by living persons. Even the provisions concerning dead bodies do not refer to the disposition and use of unclaimed corpses. The entire subject of commerce or payment for body tissue is left untouched, as is the problem of allocating tissues when there are not enough to go around. On a deeper ethical and intellectual level, the Act has been criticized for its order of priorities in designating those who can direct the use of a dead body: Section 7 gives an overriding power to the government to conduct legally required autopsies or postmortem examinations, and exercise of this power cancels out an anatomical gift. Second priority is given to the wishes of the deceased, third to the wishes of relatives, and fourth or last to the treatment of the sick. Disapproval of this sequence is based on the proposition that the highest principle of ethics, religion, medicine, and law is the saving of human life. Accordingly the right of the sick to the use of the tissues of the dead should supersede the wishes of the deceased and his family, as well as any right of the community itself (for example, when the community requires an autopsy to determine whether a crime has been committed).

As perceptive as the last criticism may be, it ignores political realities in its demand for a fundamental reassessment of our attitudes toward the dead and the living. Western society in 1968 was not ready for such a reassessment and is only now beginning to consider one. In 1968, the prospects of persuading all the states in America to enact a law that would empower medical practitioners to assume primary control over all dead bodies would probably have been nil, and it would have made little difference that the object of the law was solely to benefit the sick and the dying. The Uniform Anatomical Gift Act, despite its limitations, was a breakthrough. Its very success in gaining general acceptance and becoming law may have been due to its modest scope and similarity to a number of then-existing state laws. A more controversial measure that asserted new and widespread claims over the body could not have achieved this.

Even the powerful economic arguments put forward would not have made much difference. These blunt arguments were designed to show that it is uneconomic to allow citizens to die when they might be cured by transplant surgery and continue to contribute to the economic growth of the community, and that the costs of treating chronically ill patients far outweigh the costs of taking the tissues needed to cure them—for example, long-term dialysis costs as opposed to those of a kidney transplant. Despite their validity, such economic arguments alone are not likely to change deep-rooted human feelings, unless they are accompanied by understanding and acceptance of the worth of the change.

Now, however, more than a decade after the Uniform Anatomical Gift Act in the United States, there is evidence of changing attitudes toward the use of dead bodies throughout the Western world, particularly in Europe. Europeans are moving toward greater state power. Whether this initiative will produce an entirely new array of transplant laws remains to be seen, but it seems likely. The onus will then be put on American lawmakers to decide whether their society is ready for a

fundamentally new approach. People now have a far greater understanding of the ramifications and complexities of the removal, transplantation, and grafting of body materials from one person to another;
lawmakers are beginning to learn the facts that properly justify different attitudes toward the use of different tissues—the important distinctions between regenerative and nonregenerative tissues and paired
and unpaired organs, for example.

The 1975 Council of Europe report referred to earlier was prepared
by a committee that met in March of that year in Strasbourg. This document, dated April 23 and entitled "Ad hoc Committee to Exchange
Views and Information on Legislation in Member States Concerning
the Removal, Grafting, and Transplantation of Human Organs and
Tissues," paid attention to the removal of human tissues from both the
living and the dead. Drawing a line between "biological substances
which can regenerate" and those which cannot, the committee stated
that not one of the member countries had any specific laws covering all
aspects of the removal from living persons of regenerative tissues. A
few countries (notably Austria, Belgium, and France) regulated blood
transfusion, but that was all. As for nonregenerative body parts, the
report found that only three member states had special laws: Denmark, Italy, and Norway.

In relation to the dead, the committee reported that laws authorizing removal of body parts existed in Denmark, France, Italy, Norway,
Sweden, and the United Kingdom, and that a number of other states,
including West Germany and Belgium, were considering draft codes
on transplantation and the therapeutic use of human tissues. They and
the remainder simply relied for their existing legal rules on "general
principles of civil and criminal law and medical ethics or analogy of
the rules which govern other matters related to dead bodies."

As a result of the report, the Council of Europe prepared a model
legal code.

The model code regulates tissue removal from both the living and
the dead, and extends to all tissues of the body except embryos,
ovaries, ova, sperm, and testicles. These are obviously considered to
raise separate social, moral, and ethical problems that call for distinct
legal rules—none of which have yet been prepared. Under the code,
commerce in body materials is prohibited. To prevent international

trade in such materials, the resolution adopting the code invited member states to apply the anticommerce law to tissues that cross their boundaries and emanate from nonmember countries.*

The Council of Europe weighed the interests involved and found the balance in favor of greater community access to dead bodies and against the continued necessity to obtain the individual's consent or that of his family. Article 10 provides that tissues and organs may be removed from a dead body if there is no recorded objection by the *deceased*. No inquiry of relatives is needed. Although the code itself acknowledges that some member nations may still prefer to make a law that involves the deceased's family in the decision to remove tissue, the Council's own preference is described in the following comment in an explanatory memorandum dealing with the dead person who has said nothing:

> In such a case it has been considered that a presumed consent exists since in most states, where everybody knows that removals can take place, those who are strongly against any possible removal ought to have made it known. In adopting such a solution, *which is the one adopted by most recent legislation in Europe,* the majority of experts was essentially inspired by *the invaluable importance of substances for transplantation, the shortage of substances available and the interests of sick persons.* (Italics added.)
>
> The experts were aware that such a rule of presumed consent is for the moment and for some states a far-reaching one, since legislations or practices require an inquiry into the views of the family of the deceased person. . . .
>
> Therefore Article 10 must be considered as a long-term aim, states which are not yet ready to accept it having the possibility of providing for the inquiry in question.

If the deceased has indicated an objection to the use of his or her body for therapeutic purposes, that objection will prevail. Central to the European recommendation remains "the principle—which admits

---

* International trade in human body materials is no fiction. The purchase of large quantities of so-called Euroblood by American blood banks has been taking place for some years, and in Chapter 1 we saw evidence of the international trade in fetal tissues. In various parts of the world there is also trade in sterilized blood vessels, in placentas, and in other tissues.

of no exception—of respect for the wishes of a deceased person." The Council also urged that all member states "intensify their efforts to inform the public [of] the need and importance of donations of substances. . . ." There could be no clearer confirmation of the value of the human body.

The code's revolutionary call for greater state powers over dead bodies reflects a growing European consensus (although progress is, as might be expected, uneven) and takes a long stride away from the philosophy of the American Uniform Act, with its bias (some say excessive bias) toward the individual donor. The Uniform Act does not contemplate the use of a body without some form of consent from or on behalf of the deceased or his family. On the other hand, the practical result of the European rules will be that governments will allow a hospital to take body parts from most dead patients without any consent.

A further indication of the importance the Council of Europe attributed to therapeutic usage of human body parts was the passage of a follow-up resolution on March 14, 1979, whereby the Council recommended model laws to control and facilitate "international exchange and transportation of human substances," including for safe, speedy packaging and transport, freedom from customs formalities and taxes, elimination of the profit motive, and the exchange of information.

Has the Council of Europe model code, or its underlying philosophy, yet found expression in the laws of any member nations? The answer is strongly affirmative. In 1976, following the 1975 report but long before the perfection of the code in 1978, both the French and Portuguese enacted laws based on the new concepts of wider availability of cadaver tissues. In September 1978, a comparable bill was introduced into the West German parliament, and in December of the same year, Greece enacted a public statute. Turkey followed with a public statute in May 1979 and Spain in October 1979, while in the same month a private member's bill was given a first reading in the British Parliament.

On December 22, 1976, Law of France No. 76-1181 "concerning the removal of organs" was promulgated by the President of the Republic. It is called the Caillavet Law, after Senator Henri Caillavet, who introduced it. This law provides that immediately after death a person's organs may be removed from his body for therapeutic or scientific purposes provided that he has not specifically objected or refused during his lifetime. A hospital need not approach the deceased's family (except for the bodies of minors and mental defectives). This statute was followed, on March 31, 1978, by a decree of the Council of State prescribing the procedures whereby objection or refusal may be evidenced or recorded. The Portuguese law, which preceded the Caillavet Law by six months (June 13, 1976) and Portugal's own membership in the Council of Europe by three months, is a no-nonsense decree that bluntly authorizes removal from all dead bodies of parts "required for transplantation or other therapeutic purposes" and prohibits the supply of any information to the deceased's family of the actual use made of the tissues. Removal is not lawful if the doctors know that the deceased was opposed to the procedure of removal.

Looking eastward, comparable laws to this and the French can be found in Hungary and Czechoslovakia. In Bulgaria, there is a suggestion of a drift toward total state powers: a law of April 1976 appears on its face to authorize removal of tissues for transplantation from hospital cadavers as and when required. It contains no reference to objection by the deceased.

The Western democracies in Oceania and the Southern Hemisphere have also been active. It is interesting that the new European model code of 1978 bears many resemblances to a code drafted by the Australian Law Reform Commission that was placed before the Australian Parliament in September 1977 and first became law in Australia in December 1978. The stress in the Australian law is upon increased availability of body materials for use in treating the sick. Hospitals are empowered to remove body parts from dead patients, but only if the patient or close relatives indicate that they do not object. This is midway between the American requirement and the European recommendation. Like the Europeans, the Australians prohibit commerce in human tissues, regulate specifically the removal of tissues from the living (in particular children and incompetents), and deal with the diagnosis of death in brain-damage cases. Their law also extends beyond the use of body materials for transplantation to use for other therapy and for medical and scientific ends. Before the adoption of this entirely new statute, most of the Australian laws in this field, like the

New Zealand Act (passed in 1964), were copied or adapted from the English Act of 1961.

In the Americas, Canada first developed a uniform act in 1965, and in 1971, the Uniform Law Conference of Canada prepared another, known as the Human Tissue Gift Act. With only one exception, the Canadian provinces all adopted laws based on (or similar to) the uniform statutes. The Canadian approach, like that of the United States, is to require an actual consent to cadaver tissue removal somewhere along the route, even when there is no known objection by either the deceased or his family. In Mexico, too, under a law of August 1976, an active consent must be obtained from either the deceased or the nearest relative, except when an autopsy is legally necessary, in which case organs and tissues may be used for transplantation. Brazil also allows an exception to the necessity of obtaining a consent. When there is no known objection and no "person responsible for the cadaver," removal of parts is "permitted with the written authorization of the director of the institution where the death occurs."

Argentina introduced an unusual law in March 1977. This measure does not insist upon consent but places a positive duty upon hospitals to obtain from each patient (or relatives, if the patient is unable to understand) an indication of attitude to the removal of tissues in the event of death. The law then provides that a person *may* authorize the removal of tissues from his body after death. If he does not do so, his relatives may, and in their absence, the hospital may.

South Africa, no doubt because of its historic connection with heart transplantation, has sophisticated legislation in this field, the Anatomical Donations and Post-Mortem Examinations Act of 1970.

Most European nations that have not yet adopted the 1978 philosophy of the Council of Europe are edging toward it. Under Danish, Norwegian, Swedish, and Italian laws, there are circumstances in which tissues may be taken from dead bodies and used for therapeutic purposes if there is no known objection by the deceased or his relatives (even so, the Scandinavian laws suggest, relatives should be consulted if possible; in Italy, if a relative objects, that is the end of the matter). Finland requires either an active consent by the deceased or, if he expressed no opposition to tissue removal while alive, an indication of nonobjection by the nearest relatives.

Some advanced Western societies have held back from making any new laws at all to regulate body-part removal and the use of human tissues for therapy. Two of these are the Netherlands and West Germany. What do they do instead?

The Dutch Ministry of Public Health expressed the view in 1980 that it would be "several years before" the Netherlands Parliament would be prepared to enact a special law on transplantation, "notwithstanding the report of the Council of Europe." Even though no legislation has ever been passed in Holland, it has been found necessary to devise a set of rules. This has been done by taking the century-old existing laws on autopsy and using them "by analogy" to regulate cadaver tissue removal. In this way, accepted legal principles govern the donation of body parts by living and dead donors and by minors, and a mechanism exists for obtaining the consent of relatives.

West Germany is in a more forward position. In February 1980, the government expected that the Council of Europe–inspired Transplantation Bill which had been before the Bundestag since September 1978 would become law in due course. This did not mean that a legal vacuum existed in West Germany. By processes similar to those in the Netherlands, rules had been evolved from other existing German laws. For example, the removal from a dead person of an organ for transplantation is controlled by the "general provisions of criminal law for the protection of human life, of physical integrity and also against disturbance of the peace of the dead." According to the 1975 Council of Europe report, the "evolved" West German law provides that if no objection has been expressed by a deceased person to tissue removal, his consent may be presumed. This would obviate the need to refer to relatives for consent or nonobjection. Even more startling is the exception that in an emergency organs may be taken to save another patient's life despite any objection the deceased may have had. The German courts have devised principles that govern and control tissue donation by living adults and minors.

Belgium, too, has apparently tolerated over a long period nonlegislative rules of a peremptory nature that permit the removal of cadaver body parts without the consent of relatives in cases where no objection has been recorded by the deceased. The Paris newspaper *Le Monde* in November 1976 carried a comment by an eminent Belgian medical practitioner to the effect that a regulation along the same lines as the Caillavet Law had been introduced in the teaching hospital of the University of Ghent as early as 1965. From that date onward, the powers of tissue removal were regularly used in transplantation, and the regulation had served as a model for other Belgian teaching hospitals.

We can see from this brief historical reprise that Western social attitudes concerning the use of the body for "community" purposes have tended to take shape after a particular sequence of events. First comes a pioneering medical advance—for example, the appearance of viable kidney transplants in the 1950s, at a time when medicine developed the technical capacity to keep respiration and blood circulation going in patients who were dead because their brain function had been destroyed. The original successful heart transplants provide another example. The medical advance is followed by a period of shock during which the community—through its doctors, scientists, philosophers, theologians, and lawmakers—grapple with its wider significance. On balance, if it is believed that society will derive a benefit, the burden of ensuring community acceptance is transferred to the lawmakers, who must devise laws that will enable the medical and other necessary routines to be carried out lawfully, framing them so as not to allow the procedures to trespass beyond the limits of community acceptance.

Although this progression is not confined to transplantation or even to medical subjects, it has special importance for transplant surgery because the laws on this subject authorize drastic invasions of the human body. A society that seeks access to the bodies of its citizens for the removal of their contents obviously enters a highly sensitive personal area.

Uses for body materials have grown and diversified so rapidly in the past thirty years that legal regulation has become essential. The medical procedures are too novel to be left in legal limbo, for fear of outraging public opinion and in some cases throwing those involved into direct conflict with existing laws. The deliberate employment of the legal system to regulate the removal of body parts has therefore served a number of purposes. Creating a public law is a sign of official approval; it gives the community an authorized system and at the same time an opportunity to digest the rules and react to them. It is well-known that a law that arouses continued public hostility will not survive in the long run. As community attitudes develop and are communicated to the politicians, they in turn can synthesize public feeling, scientific progress, and community benefit.

In the matter of the human body and its rights—where the benefit sought may be achieved only by disturbing very basic human sentiments about the dead, or by arousing sleeping folk memories of slavery, or by imposing totalitarian control over the individual—it is no wonder that progress has been uneven. Some states have taken one steady legal step forward after another, some have taken one step forward and have stood still in fright, while others have done nothing at all. However, the price of doing nothing is the inability safely, in a legal sense, to take and use human tissues to help the sick.

CHAPTER 4

The Acquisition of Organs and Tissues: Contribution or Confiscation?

The principal means whereby Western society has been led to accept the use of human "spare parts" to treat the sick has, as we have seen, been the enactment of public statutes. What are the underlying principles of these laws and decrees? What direction are they taking?

One of the main purposes of any public law will be the acquisition of enough body materials to satisfy community requirements. We have already seen that demand far outstrips supply and that medical capacity is not always fully used. Long-standing, acute shortages of organs are themselves an indication that existing laws are inadequate. The question is how to increase supply and give the sick and dying the tissues that can cure them.

Before any community looks at ways to increase tissue supply, it should ask itself whether there is, in truth, a sufficiency of suitable "spare parts." Assuming ideal conditions, whatever they may be, are there enough to go round? Plainly it would be reprehensible to create laws that impinge on individual autonomy and freedom if they had no prospect of ever achieving their purpose. Not surprisingly, little factual or scientifically acquired information is available to answer the question satisfactorily.

Some facts were provided in an unusual report published in the *Journal of the American Medical Association* in October 1972. A specialist team from the Medical College of Wisconsin carried out a retrospective study of all deaths that occured during one year in Milwaukee County, Wisconsin, which then had a population of 1,072,100, with the purpose of determining whether enough suitable kidneys and livers were available to satisfy the needs of prospective transplant patients. Unhappily, the transplants were not performed, because neither the means nor the facilities for obtaining and transplanting the organs existed. The study showed that the number of potentially suitable organs was greater than the number needed.

The total number of deaths in the population in 1966 was 10,494 (about 1 per 100), but the study group examined only the records of those who died in a selected number of hospitals, chosen because of their facilities and convenience for transplant surgery. An interesting, perhaps surprising finding was that 65 percent of all deaths, 6,681, occurred in those hospitals. In order to determine how many of the dead hospital patients would have provided suitable organs, the team laid down strict criteria and required that full medical records be available on each patient, showing medical history, freedom from disease and defect, and details of the relevant organs. From the information available, it was decided that 460 cadaver livers and 60 kidneys were suitable for transplant. In calculating the number of putative recipients, the medical team considered the population at large, and obtained details of all registered deaths in Milwaukee County from kidney and liver complaints of the kinds that are considered medically capable of being cured by transplant; they added the number of kidney patients on dialysis. The total number of people who could have benefited or been cured by transplants were 117 liver patients and 56 kidney patients.

Proceeding from the particular to the general, the authors cited another United States analysis that had shown "that under ideal logistical conditions the national demand for kidney and liver transplantation could be entirely supplied by cadaver donors." The Milwaukee team drew attention on more than one occasion to the scarcity of tissues: "Moreover most transplant centers are still faced with a lack of organ donors for the ever-growing list of recipients and can foresee only an increase in this disproportion as future advances in transplantation develop." An expert in kidney disease from the University of California Medical Center said in 1972 that the number of kidney donor cards which statistically should have been carried by voluntary donors, in order to produce from hospital deaths the number of kidneys that could be transplanted with existing facilities (10,000), was not the 10 million that had in fact been issued, but 100 million.

How should organs and tissues be obtained? Should they always be voluntarily given or are there circumstances in which they may be *taken* without consent? Remember that we are talking of both the dead and the living as sources of tissue. Medicine is happy to accept the tissues of both, although obviously it would find it far more convenient to obtain all body materials from living persons. Unfortunately this convenience would be enjoyed by all concerned except the living donor, so it is sensible first to look at the dead human body as the more likely source.

Probably the oldest method is that which allows a person to make a gift by his will. The use of written documents, including wills, for donating one's body for anatomic study is long-established, having been made lawful by the English Anatomy Act of 1832. However, the credit for being first may belong to Massachusetts, which passed an anatomy statute in 1831, anticipating, according to one writer, "all sister states and England as well." Many laws specifically envisage that the desire to donate tissue will be signified in a will. But a will is not really a practical vehicle for giving the most-needed tissues for transplant, for its validity and effect are not formally established until it is officially proved or probated—often a considerable time after death, in some cases months. Organs for transplant are needed virtually at the time of death or within hours thereafter, while a body given for anatomy may not be needed for weeks or months, and can retain its usefulness indefinitely, if properly stored. The employment of wills for gifts of tissue for transplant and general therapy should not be discouraged, but their limitations ought to be understood. Some statutes, such as the Canadian and United States Uniform Acts, contain specific wording which allows a tissue gift by will to take effect upon death without awaiting the formal grant of probate, but even this sort of provision will not be practicable if organs are needed within minutes or hours. The gift will be impossible to accept if the will, as wills tend to be, reposes in a bank vault or a lawyer's office safe and cannot be read for a day or two after the person's death. The hospital surgeon would need to satisfy himself immediately that he is acting within the law. Transplant statutes that permit gifts by will also normally envisage tissue gifts by less formal means—for example, a simple written document or card, or oral declaration in front of witnesses.

The popular system of donor cards has the merit of simplicity and portability. A typical example is the uniform card developed in the United States following the Uniform Anatomical Gift Act of 1968. This card, which can easily be carried in a pocket or wallet, states in simple words the donor's desire to make an anatomical gift to take effect upon death. The operative words are:

. . . .

                                                   (name of donor)
in the hope that I may help others I hereby make this anatomical gift, if medically acceptable, to take effect upon my death. The words and marks below indicate my desires.

I give: (a) _____ any needed organs or parts.

           (b) _____ only the following organs or parts.

. . . .

                                (specify the organ[s] or part[s])
for the purpose of transplantation, therapy, medical research or education.

           (c) _____ my body for anatomical study if needed.

Limitations or special wishes if any

. . . .

On the reverse side the card contains provisions for signature, witnessing, and personal details. Similar cards are used in other parts of the world, including Australia, Canada, and England.

Unfortunately there is little concrete evidence in any country that these cards have substantially increased the availability of kidneys and tissues for transplant. It is not uncommon for leading renal surgeons in communities that have distributed many donor cards to say that they have never encountered a dead kidney donor who carried one. Still, it would be wrong to discourage the use of cards. Community awareness of the desirability of donation is very important, and the spread of these simple documents is an easy way to increase awareness. They have an important educational function and are strongly supported by governments and doctors. On occasions the fact that a dead donor was carrying a card receives widespread publicity. A sixteen-year-old girl who was fatally injured in January 1980 in Leicestershire, England, when her motorcycle collided with a truck was carrying a donor card, and after death was pronounced, her heart, kidneys, and corneas were donated for transplantation with the agreement of her father. The surgeon in charge of the case said that these lifesaving and sight-restoring gifts were greatly facilitated by the card because it made it easier to raise the subject with the donor's parents, and presumably made the parental decision easier too.

In many places, organ donation has received direct government support in association with the issue of driver's licenses. This is a very practical procedure, because automobile accident victims are a major source of suitable transplant material. Britain, Canada, and the United States all use it. In Australia in 1975, a national poll indicated that 85 percent of all citizens favored the use of driver's license stickers, attached to the license documents, which bear the words "I am a kidney donor." The government encloses a sticker in the envelope containing every renewed driver's license. It is up to the licensee to decide whether to put it on the license.

In February 1978, *The New York Times* reported that all fifty states now use some licensing or registration system to distribute organ and tissue donor cards or stickers, or some other donor indication. The director of the Kidney Disease Institute of the New York State Department of Health said that New York State distributes donor cards with renewed driver's licenses, that many doctors and lawyers now keep stocks of donor cards in their offices, and that "eventually the donation of organs and body parts will be as commonplace as donating blood."

Many other methods of signifying a wish to be a donor of tissue upon death have been practised and suggested. Some favour wearable objects such as pendants, bracelets, key rings, and chains with the relevant information engraved on them. For those who prefer more privacy, there are systems such as "Medic Alert," which involves wearing a wrist bracelet to which is attached a metal disk bearing an engraved number plus a telephone number; in the event of injury or death, information concerning the wearer and his wishes may be obtained by telephoning the number, which is a twenty-four-hour computerized data bank. This system is used in a number of countries, and the Living Bank in Houston, Texas, operates a similar facility.

A Dutch system involves a wearable article, the "S.O.S.-Talisman," which looks like a small wristwatch. It unscrews to reveal a strip of paper on which provision is made for personal information to be recorded. The S.O.S.-Talisman may be worn on a chain around the neck, as a bracelet on the wrist, or on a watchband (with or without a watch), and it is made of inexpensive silver or gold metal that can withstand extremely high temperatures. With all objects of this kind there are prerequisites to usefulness, such as low price, ready availability to the public, ease of recording one's wishes, and the likelihood of the wishes being communicated to an appropriate recipient in the event of death.

Other methods, more suggested than practised, include the use of body marks or tattoos which would show information on the skin of a donor, for example, under the arm. Such marks would have the advantage of ready access to information and would preclude the possibility of loss which exists with cards and wearable objects. On the debit side are the difficulties of having the information recorded in the first place, and of erasing it if a donor changes his mind. Some people also object to disfigurement of the body as a matter of moral or religious principle. Indeed, not everybody approves of the use of wearable objects to signify willingness to give away body parts. One well-known writer has flatly said, "Surely there is something macabre about a society in which people go around wearing organ-donor bracelets."

Behind these *modi operandi* are many local, national, and international organizations that work to persuade the public to give tissues, including transplant societies, kidney associations, pituitary agencies, tissue banks, health foundations, and blood banks. Other organizations have been created solely to facilitate the obtaining of body parts for therapy, such as the Living Bank of Houston, Texas, or to provide continuous transplant information, such as the Organ Transplant Registry of the American College of Surgeons and the National Institutes of Health, which operated in Chicago until 1977, when it handed over its functions to other groups. The Living Bank is a nonprofit service organization that obtains promises from people to give their body parts upon death for medical and scientific purposes. It not only encourages donors but acts as a repository of information for doctors and hospitals as well.

Service organizations such as Rotary Clubs, Apex Clubs, and Lions Clubs are frequent instigators of local community movements to encourage organ donation, and have set up eye banks and organ banks in the United States, Japan, and other countries. There are eye banks in many parts of the world, and more than fifty in the United States.

However, the use of the word "bank" in this context can be misleading. A tissue bank is not necessarily a place where frozen body parts are held until someone needs them, though a limited number of human tissues can be stored for quite a long time—for example, frozen red blood cells, plasma, pieces of bone, skin, and tendon. Organs for transplant cannot be held for more than a very short period after removal and cooling (to around 4° C.)—four hours for hearts, twelve for livers, and two days for kidneys (although kidneys have been kept for up to four days before transplant). As for eyes, although long-term storage of corneas has been made possible by techniques such as freeze-drying and desiccation by glycerine, an "eye bank" may be no more than a place where eyes or corneas are held briefly, or where information about available eyes or corneas is recorded.

"Tissue banks" today are often directed more toward acquiring information that will enable donors and donees to be located and "put together" when needed than toward accumulating tissue. Information is obtained from living volunteers, from patients upon admission to the hospital, and from direct public soliciting, particularly concerning regenerative tissues such as bone marrow and blood. Some computerized services not only cover more than one kind of tissue but extend far beyond national boundaries. They produce instant matching data from stored details of waiting patients when an organ becomes available for transplant. The United Kingdom Transplant Service (UKTS) is a good illustration. It is associated with the Eurotransplant Foundation and the European Dialysis and Transplant Association, whose records include patients in Israel, Cyprus, and Finland. The transport of human tissues and organs from one nation to another has, in fact, become highly organized and receives international legal attention. Similar exchanges are practised between Australia and New Zealand. Public polls have frequently shown that substantial majorities of the population are willing to donate organs—70 percent of American adults under a Gallup Poll reported in *The New York Times* in January 1968, 82 percent of Australian adults in a Gallup Poll reported in October 1975, and 57 percent and 69 percent, respectively, in 1978 in Britain in a National Opinion Poll and a Gallup Poll.

An illuminating study of the characteristics of voluntary human tissue donation based on work done by a research team at the University of Minnesota was published in 1974. This followed a newspaper, radio, and television campaign launched in Minneapolis–St. Paul by a university hospital to persuade people to commit themselves to giving body parts for transplantation after death by applying for and signing donor cards. Out of a population of 566,000 over the age of fifteen, the campaign yielded 3,100 signed cards within six months.

Then the university team compared the attitudes of eighty donors with those of a control group, and also made statistical comparisons with the population at large. The study disclosed much that was significant. For a start, 95 percent of the donors had at least four years of high-school education as opposed to 62 percent in the general population, and 36 percent were college graduates compared with 11 to 12 percent in the general population. Other studies have supported this finding that better-educated persons have a more approving attitude toward science and medicine, and are more favorably disposed to organ donation. The donors also tended to be younger (some 68 percent were between the ages of fifteen and thirty-nine) and were more likely to be female (72 percent) than the general population, in which 48 percent were between fifteen and thirty-nine, and 54 percent were female. The donors showed less traditionalism and less orientation toward religion than the control group. Their answers were especially instructive on the subject of attitudes toward death. Questions were asked about belief in life after death, fear of death and the hereafter, belief in resurrection, and whether the human body is sacred after death. Of the control group, 36 percent believed that the body is sacred after death, while only 14 percent of the donors felt that way. An overwhelming 93 percent of the donors believed that "a person's body can be put to better use than simply occupying a cemetery plot."

A puzzling aspect of voluntary donation is the gap between the numbers of people who appear to be happy to give away their organs and tissues for therapeutic purposes at the time of death, and the low numbers of organs actually received. Many people, including prominent transplant surgeons, place the blame for this failure at the door of the medical profession itself. Professor Roy Calne has publicly commented on the reluctance of many doctors to assist in the transplant process. The reasons range from their reluctance to approach grieving relations, to an inability to "admit failure" when a patient is lost after strenuous efforts to cure him. A well-known English medical writer says that this attitude in the medical profession "blocks the gift impulse in some way and prevents the tissue from getting to those who need it."

It is abundantly obvious that public education is needed. The more the community realizes the value of giving tissue, the more the supply is likely to increase. Under any system that relies on consent, continual encouragement to donate is required—through government and private publicity programs, the media, schools and educational institutions, and the support of all acceptable means of indicating the wish to give. An impressive example is supplied by the Eye Bank for Sight Restoration in New York City, which will supply speakers and donor cards upon request to potential donors, schools, and civic groups. One ethicist, after an examination of human tissue shortages, called for "a firm campaign [to be] waged against our shameful waste of human tissue . . . by ministers, chaplains, doctors (especially general practitioners), teachers, public health officials, publicists, and many others whose support of medicine . . . can still overcome the mini-morality of body taboo."

It is not difficult to see why many people favour the idea of cadavers being made automatically available for tissue removal unless a prior objection has been recorded. Reliance on active consent, or the spread of altruism, simply has not achieved the needed results. Yet the wisest procedure may not be to reject voluntary donation and change to another system altogether. The answer may lie, as do the answers to many human problems, in compromise and the use of the best features of all available possibilities.

At one extreme stand those who insist that no body part should ever be removed without the prior consent of the person from whose body it is taken. If the patient said nothing while alive, then the corpse must be left alone. At the other extreme are those who say that the community should have the power, on behalf of the sick, to take any material from a corpse regardless of the wishes of the deceased or his family. "When you're dead you're dead!" says this school. In the first instance, the principle of personal autonomy reigns supreme, while in the second, the community lays claim to the individual's body and asserts a kind of ownership which takes priority over his own wishes.

There are already, in medical fields other than transplantation, established laws of the second kind above that assert dominion over the dead. If the body is thought of as a quarry, then there are hunters stalking it, and some are heavily armed with compulsory powers. Fortunately the quantity of body materials they require is not large in absolute terms, but it may come as a surprise that there are such laws at all. Some of these laws brush aside all contrary wishes and objections of the deceased and his family. When these are viewed cumulatively, and then added to the tissue-removal laws which now serve general therapy and transplantation, the need to protect the interests of the individual becomes even more obvious.

The laws in question are mainly those which regulate official autopsies, anatomical examination, and blood transfusion. Any statute on these subjects is likely to provide for compulsory procedures involving removal, implant, or study of tissues.

Autopsies by coroners or designated state officials are well-known, but they should not be confused with normal hospital autopsies, which cannot be carried out without the consent of the dead person or the relatives. As a rule, official autopsies are legally required in cases of homicide, suicide, or death by other violent or unusual means, and neither the deceased's wishes nor his relatives' can prevent them. The public's interest in ascertaining the cause of death overrides private objections to the violation of the body.

Normally, the power to perform an official autopsy will take precedence over the power to remove tissue for transplant or therapy, or even cancel a gift of tissue. The United States, Canada, England, and many other countries have specific provisions to this effect. Critics have objected to this "right of eminent domain" on the ground that the purpose of a gift of an organ or body part for transplant has a higher value than the public's "right to know." On the other hand, some coroners take a strict view of their statutory powers. In the case of the English girl who was carrying a donor card when she suffered fatal injuries in January 1980, the local coroner claimed he had not been correctly consulted before her heart was removed. In its account of the coroner's inquest in late February, *The Times* of London revealed that the coroner had ordered an investigation by senior police officers into the facts surrounding the surgical removal of the girl's kidneys, eyes, and heart (they failed to find sufficient evidence for a prosecution). "All I was asked for was a kidney, which I granted," the coroner said. Despite the willing gift of a number of organs by the girl's parents and their successful use by transplant surgeons (a spokesman for the surgeons said, "The lives of three people have been enriched and lengthened by the operations"), the coroner was prepared to launch criminal proceedings against the doctors for failure to observe the letter of the law.

In a number of countries, autopsy laws have been extended to permit additional use of body materials removed during coronial autopsies—a significant development, bearing in mind the compulsory nature of autopsy statutes. In some places, removed materials may be used only for specific, named purposes, but in others, permitted usage extends to transplantation and general therapy.

These new autopsy laws raise real problems for lawmakers, ethicists, and moralists. In a manner of speaking, they provide that two wrongs can make a right. The problems begin with the gruesome truth that a properly conducted full autopsy will involve the draining away of all blood and body fluids, and the removal of all organs, glands, and the brain. Once the cause of death is ascertained, the body will then be restored to a normal appearance, so far as possible, and handed over to relatives for interment. Frequently, severed parts are carefully placed in a body cavity such as the abdominal cavity, and all incisions are stitched up. But organs and glands cannot be put back in their original positions and there is no point in giving further thought to body fluids.

It is at this stage that the dilemmas arise. Should tissues with therapeutic value be replaced along with the other parts, or should they be kept for treating the sick, for research, or for medical education?

Consider the pituitary gland. Located at the back of the skull, this pea-sized gland has immense therapeutic value, so much so that a number of nations have programs to gather pituitaries and make hormone extracts from them. In the course of a normal autopsy the pituitary gland is removed and examined. It cannot be put back in the same place because of the damage done to its connecting tissue by severance. Because it is so small, it would more likely be treated as waste after examination and thrown away than kept for lodgement in a body cavity with other severed parts.

The United States National Pituitary Agency has for some years been collecting upward of 80,000 pituitaries a year and constantly seeks greater supplies. We have already seen that under the Australian program some 300 children who would otherwise have remained dwarfs and 250 women who would have remained infertile were successfully treated with pituitary extracts. Increased availability of pituitaries is enabling the treatment of greater numbers. The number of pituitaries collected in Australia rose from about 7,000 in 1970 to 12,000 in 1977. As a result of the public benefit conferred by these programs, some societies have underpinned their schemes by supporting legislation. In Italy, California, and Washington State, special laws have been passed that facilitate pituitary-gland collection. The obvious source of continuous supply is the coroner. Accordingly, these new laws make it lawful to hold pituitaries removed during official autopsy for public health purposes.

Some societies have taken this kind of law much further, with statutes that allow retention for therapeutic use of *any* tissues normally removed during autopsy, e.g. a Hawaiian law of 1968, a South African law of 1970, a law of Oregon of 1976, and an Australian law of 1978.

Traditional coroners' statutes make it clear that the sole aim of coronial autopsy is to establish the cause of death. It follows that it could be unlawful to use removed body parts for any other purpose, no matter how commendable. Yet it would surely be absurd if this tiny severed gland, obtained after a gross interference with the physical integrity of a dead body, could not lawfully be used for curing sickness but must instead be destroyed or sewn up in some other part of the body. The new laws are both practical and humane. They say, in effect, that medical advances now enable the public to benefit from many body tissues routinely removed in coroners' autopsies and that clearly these tissues ought to be used for purposes of public health. The laws are activated only after the coroner's knife has produced tissue in usable form. It would be a human and economic waste to forbid this use and to compel destruction of the tissue.

This point of view is persuasive, and my own opinion is that such laws should be encouraged. But a body of strong opposing opinion exists. Opponents begin with the proposition that official autopsy is justifiable only because the public has an interest in knowing the cause of unusual death, and that noncoronial autopsies can be lawfully performed only with the consent of the deceased or relatives. The fact that autopsy seriously interferes with the human body is considered to be a necessary evil. Once the procedures needed to determine the cause of death have been followed, everything possible should be done to restore the status quo. This approach rejects the argument that says: Now that you have gone this far, you might as well make some use of the available materials for treating the sick or defective instead of destroying them or allowing them to rot. It says instead: You shouldn't compound the affront to bodily integrity by committing an additional offense of the same nature. You have gone so far but should go no farther. If coroners now take some body parts, where will they stop when uses become more widespread?

Not only autopsy law has cast envious eyes upon the body and its contents. Anatomy rules of the nineteenth century sometimes went beyond the point of waiting for the donation of bodies, and reached out and grabbed them. In Europe and elsewhere, anatomy laws are still in force that survive from the days of charity hospitals and public institutions, when the dead body of an inmate was regarded as a fair exchange for the cost of maintaining him while alive. Blood transfusion is also a subject on which any community is likely to possess a legal code. Some, such as the United States and Australia, provide for compulsory blood transfusions to children in life-or-death cases where parents or guardians refuse on religious grounds or the like. Compulsory removal of blood is not yet known to the law in any Western country.

Another means of nonconsensual or even peremptory access to the body has surfaced in some recent transplant statutes through deliberately drafted "loopholes." Thus, the 1976 Mexican law provides that bodies not claimed within seventy-two hours of death may be used for research without consent. People who live in a community whose transplant laws are founded upon the principle of voluntary donation may be surprised to realize that extensive compulsory autopsies are performed regularly and that compulsory blood transfusion may be lawful. In order to achieve a proper perspective, we should correlate the various statutes concerned to ensure the availability of bodies—whether for autopsy, anatomy, blood transfusion, research, education, transplantation, or general therapy.

As for compulsory violations of bodily integrity, a positive array of statutory licenses authorizes it. In any society, one is likely to find that criminal law enforcement authorities may take samples of hair, blood, or semen, or that public health purposes may require compulsory vaccination or quarantine. Body searches by customs officers are well-known. Blood tests are a prerequisite to marriage in some jurisdictions. Military service, compulsory schooling, the duty to assist in law enforcement, and the duty to help during natural calamities are also familiar legal requirements.

In England, the expressions "opting in" and "contracting in" and "opting out" and "contracting out" are now familiar. "Contracting in" or "opting in" means that a person's dead body should be left untouched unless he or she has taken a positive step to make a gift of it: the body cannot be used for tissue removal unless the person "contracts in" by affirmatively handing it over. Under the opposite concept, all corpses would be available to the community unless the deceased has "contracted out" or "opted out," recording a specific objection during his or her lifetime; in other words, you must contract *out* of the state's claim upon your body, and remain *in* until you choose to go *out*. It will be noted that "contracting out" envisages that the individual will have the ability to veto or neutralize state power over his or her corpse. In this sense, it is not so extreme a principle as the one described earlier under which the state would have control over all corpses irrespective of the deceased's wishes. Nowhere, so far as can be ascertained, has a government yet legislated to take a dead body in the face of an objection by the deceased; an objection always prevails.

As might be imagined, the laws of Western nations do not always fall neatly into one of these two categories. Many statutes allow a consent to tissue removal to be given by the family of a dead patient who has said nothing one way or the other. Thus empowering the family instead of insisting on the deceased's own consent is one method of keeping reasonable faith with the "consent" or "contracting in" principle while enlarging the numbers of body parts likely to be made available. It is now normal procedure for hospitals that need an organ to ask permission of the close relatives of a dead patient (close relatives are typically defined as the spouse, adult children, parents, and adult brothers and sisters).

Another relaxation of the "consent" or "contracting in" principle has been the use of "nonobjection" in some legal codes. Many doctors have noticed that relatives of a deceased patient will often hesitate when asked to say yes to the removal of organs but will be quite content to say they "do not object." This may sound like a quibble, and perhaps it is, but the "nonobjection" concept is regarded by many who have used it as carrying its own justification when it produces the desired result.

Generally speaking, Western laws as we have seen are still consent-based. They require that tissue removal must be supported by the express consent or wish of the deceased or, if the deceased gave no consent and had no known objection during his lifetime, by the express consent or wish of a close relative. Some nations have gone farther, and will permit removal when the deceased has neither consented nor objected and when there is no objection by close relatives, provided attempts have been made to get in touch with them. This could be said to strain the "contracting in" or "consent" concept to its limits. The requirement that doctors communicate with close relatives or try to do so is the last restraint between consent systems and systems under which there exist no effective barriers to the unrestricted use of dead bodies for community purposes.

Most people never think of the useful purposes their bodies may serve after death. Without widespread publicity and public instruction, it follows that few will record their objections to the removal of tissues. If the law gives to a hospital the power to remove body parts in the absence of any known objection by the dead person and without any duty to make inquiries of relatives, that power will confer, for all practical purposes, unlimited access to the corpses of dead patients. A legal obligation to make inquiries of relatives, even if such inquiries are only intended to produce a statement of no objection, clearly acts as a brake.

What do all these laws and developments, particularly the latest ones, portend? Is the human body to become simply a bag of parts that can be removed as required by Big Brother? If it were suggested outright that as a routine procedure all dead bodies should be taken for a short period by the state, drained of all body fluids, subjected to removal of the brain and all organs and glands, and then handed to the family for burial, nobody would be surprised if the public reacted with horror, anger, resentment, and resistance. Yet that is normal practice with official autopsies and there is no widespread horror or resistance. Many politicians and lawmakers have simply failed to look squarely at the truth. This no doubt accounts for the tentative quality of some recent transplant statutes, for example, those which allow tissue removal from unclaimed bodies. This sort of provision simply fiddles with the subject of compulsion, using back-door methods. If the 1976 French statute is an exemplar, the prospects are diminishing for personal autonomy or for the continuance of the principle of the human body as sacrosanct, and increasing for treatment of the human body as community property.

The Caillavet Law was proudly received in France. The newspaper *Le Figaro* described it as a "decisive stage for medicine, innocuous on the face of it, but one of a series of upheavals that over a quarter of a century has been transforming society as surely as a war." The other six (if one includes the Turkish and Portuguese) similarly oriented statutes enacted by or introduced into West European parliaments by the end of 1979, pursuant to the contracting-out recommendations of the Council of Europe, give reason to suggest that other nations will follow. A number that have kept the consent requirement have adjusted their practices and their laws so as to relax the strictness of pure contracting in. Some have gone so far that any greater relaxation will involve stepping across the boundary into the transplant territory now inhabited by France. There seems little doubt that the future course in the West will be for society to claim greater rights over the dead human body.

Under these circumstances, we must ascertain, in our own interest, what medicine and society are likely to want from us as individuals. We should consider whether these new laws are only the latest stop-over in an unfinished journey to a destination where all dead bodies will be *compulsorily* available to the state. At that point, communities may well cease to bury the dead as a routine matter, and instead use them to fill organ and tissue banks with stored human spare parts for the repair and renewal of the living. This may seem unlikely, yet it would be rash to insist that it will not come about.

Other considerations are likely to postpone, but not prevent, the arrival of this Brave New World. Assuming for the moment that the Caillavet Law makes sufficient kidneys available in France for waiting recipients, does the French nation have the hospitals, doctors, equipment, and money to transplant all those kidneys? Discussing this in February 1977, a contributor to the *British Medical Journal* wrote that 2,000 kidneys had been transplanted in France since kidney transplants began in the 1950s, but that about 1,800 French patients were then in need of a transplanted kidney and the annual transplant rate was only 350. With greater availability of kidneys because of improved surgical techniques and the new law, it was likely that demand would also increase the need for facilities, staff, and money. Similar results must be contemplated with transplants of hearts, livers, corneas, glands, and other tissues.

As an altogether separate matter, we must never forget the possibility of the growth of community claims upon the living, for it is not only the *dead* human body that is proving to be irresistibly attractive. Some transplant laws do not deal with removal of tissues from the living, but others do. Still others deal in partial fashion with live donors, such as Canada's Uniform Human Tissue Gift Act, which regulates removal of nonregenerative but not regenerative tissue from living persons.

This lack of legal evenness may be due to the opinion held in many countries that when it comes to giving away parts of their own bodies, living adults do not need close legal protection or supervision. It is hard to imagine a normal adult actively attempting to part with a vital unpaired organ such as his heart or liver. In addition, doctors can normally be relied on not to participate in foolish or dangerous operations.

Yet the latest statutes give specific attention to the removal of tissue from living persons, such as the French, the Australian, and the model rules of the Council of Europe, which all insist that free consent is essential before any removal can occur. As far as can be ascertained, all existing parliamentary laws on tissue removal from living adults insist on free consent beforehand. How could any law provide otherwise? Surely the contrary law would be a gross invasion of human rights, unacceptable to all reasonable men. Perhaps so. Perhaps some might say that it is unthinkable that any government would ever pass or permit a law allowing compulsory access to the body of a living person for the removal of his or her body parts.

Yet these idealistic thoughts do not entirely accord with reality. Already the unthinkable has begun to happen—for the best of reasons and in the interests of life and health.

For almost forty years, American judges have been building from lawsuits a substantial body of legal principles that permit the removal of human body parts from living persons who have not given, or have been incapable of giving, their own direct consents. The subjects have normally been the most helpless of all, namely, children and incompetents. But recently there have been signs in the United States of legal demands for compulsory access to the bodies of normal living adults.

Many of us know the price of liberty, but obviously many do not. If we value our personal autonomy and bodily integrity, we must at the very least try to understand how they may or could be undermined. In the present context, the problem is not the motive of those who would assert rights or claims to our bodies. The highest motive will, typically, be demonstrated—preservation of life and health and the cure of the sick. The need is to recognize danger when it is not intended, and to bring to bear a clear eye and an inquiring mind on proposals which directly affect our persons and liberties. Any new law that confers power over the human body, whether made by a legislature or by a court, has such an aspect and should give us pause.

One would not normally regard transplantation of human tissues or the preparation of therapeutic extracts as a potential source of prejudice to individual freedom—quite the opposite. Yet modern medical and scientific miracles must not be allowed to induce social myopia. Enthusiasm and the desire to promote health and improve the human capacity to enjoy life will call in aid the law and lawmakers. Paradoxically, if the resulting laws follow the direction of the latest Western statutes without a deep understanding on the part of the drafters of all that is at stake, the individual will be endangered. His body could become the domain of other men. It could become community property.

CHAPTER 5

U.S. Courts and Living Donors

I.

The United States has never had any national or uniform legislation dealing with live donors of human tissue. The fashioning of the necessary legal principles has been left to the courts, as it has in Britain and West Germany. Most Western nations lacked statutory rules on live transplantation until the late 1970s. It is only since 1975 that public laws directly regulating tissue and organ removal from living persons have begun to appear in any numbers. At the beginning of that year, three Western European nations possessed such laws, and it is doubtful whether any existed anywhere prior to 1966 (blood transfusion is excepted from these comments).

America's experience is unusual because of its historic substantial use of live kidney donors. From 1951 to 1972, almost half of all kidneys transplanted in the United States came from living donors (compared with 21 percent in Europe and under 2 percent in Australia). The habit of using live donors has resulted in litigation from which the American courts since World War II have fashioned an array of legal principles that some would call remarkable and others astonishing.

In the early days of World War II, John Bonner learned that his cousin Clara Howard was to be brought to the Episcopal Hospital in Washington, D.C., for further medical treatment. John, who lived with his mother in the city of Washington, was fifteen years old and a student in junior high school. Clara had been in an accident some time earlier, in which she had suffered such dreadful burns that she had become a hopeless cripple. Her mother decided to bring her from their home in North Carolina in the hope that the plastic surgeons at the charity clinic of the Episcopal Hospital could do something. Dr. R. E. Moran, a physician specializing in plastic surgery, examined Clara and advised that she could be helped by skin grafting, provided that the blood of the donor matched hers.

Clara's mother tried several times to fulfil this requirement but could not find a willing donor with matching blood. She then thought of her nephew John Bonner. Because his mother was sick at that time, she spoke directly to the boy and persuaded him to accompany her to the hospital for the purpose of undergoing a blood test. His blood matched Clara's, and Dr. Moran decided that John would be suitable as a donor of skin. John was asked if he was willing to undergo surgery for that purpose. He readily agreed, but did not discuss it with his sick mother, who knew nothing about the blood tests or the operation until after they took place.

When the operation was completed, Bonner was immediately sent home, having been told he must return for more surgery to complete Clara's treatment. He was not informed of the nature or extent of the surgery, which had involved the side of his torso. At home, all John said to his mother was that he was going to the hospital to have his side "fixed up."

When John reentered the hospital, he was kept there for two months, and subjected to a series of operations involving, in the words of the court, "anaesthesia, bloodletting, and the removal of skin from his body, with at least some permanent marks of disfigurement." What actually happened was that a tube of flesh was cut and formed from his armpit to his waistline, and at a chosen time one end of this tube was attached to Clara in the hope that it would "take" and eventually give her substantial physical relief.

The experiment failed, largely because of inadequate blood circulation through the tube of John Bonner's flesh. Eventually the tube had to be severed, but not until after the boy himself had lost a lot of blood and received blood transfusions. John's pain and heroism were discovered by the press during his sojourn in hospital, and amid widespread publicity a public collection of money was taken up for his subsequent education. It was during this period that his mother learned what he was doing.

At a later stage, John Bonner sued Dr. Moran for damages for assault and battery. The case reached a federal court of appeals in December 1941 and was argued before three judges. The basis of Bonner's complaint was the proposition that surgery is lawful only when it is for the patient's benefit and when the patient has given an express or implied consent. To put it another way, a surgeon is liable to pay damages if an operation is unauthorized. Bonner's consent, as a fifteen-year-old minor, was no consent at all. Because he lacked the capacity to appreciate what he was doing, an effective consent could have been given only by a parent. That consent was never sought and never given. Therefore, the surgery was unlawful, an assault and battery, and the surgeon was legally liable.

The appeals court concluded that "generally speaking . . . a surgeon has no legal right to operate upon a child without the consent of his parents or guardian." It was prepared to concede exceptions in emergencies, and in cases where the minor is close to the age of consent. (Other exceptions and changes have been made in the general rule since 1941 and are briefly described later.) But in the present case, neither of these exceptions applied, and worse, the surgery had nothing at all to do with the well-being of John Bonner. "Here the operation was entirely for the benefit of another and involved the sacrifice . . . of fully two months of schooling, in addition to serious physical pain and possible results affecting his future life."

Dr. Moran admitted in evidence that he did not explain to John or to anybody else the nature or extent of the proposed operation. However, the jury in the lower court had taken the view that his actions were dictated wholly by humane and charitable motives. By way of further mitigation the appeals court observed that the doctor, who had charged no fee at all for his work, gave his skill and professional services solely to alleviate Clara's pain and suffering.

Despite Dr. Moran's good intentions, the appeals court judges unanimously decided that John's mother's consent should have been obtained and that the doctor was wrong in carrying out the surgery without it. He lost the case, for the court recoiled strongly from the evidence of extensive bodily invasion of "an immature boy" in circumstances that required, but entirely lacked, a decision from a mature mind, well-informed of what was involved.

No doubt the decision was the right one on the facts before the court, but a number of questions remain unanswered, particularly concerning the mother's circumstances and how she remained ignorant of the facts for so long. However, if one accepts the decision as right without question, one can then reflect on its most significant message, which is that if John Bonner's mother had given her consent, he could *legally* have been subjected to the same surgical procedures. What was important to the judges was not the surgery but the consent. *Bonner* v *Moran* is a milestone for a number of reasons, not least of which is the fact that it concerns surgical transplantation of human tissue taken from the body of a person lacking "legal capacity." Together with two other well-known American cases, it leads the way to one of the most debatable lines of lawsuits and judicial decisions in the history of jurisprudence. Each of the two other cases contains memorable judicial statements, the first on the inviolability of the human body and the second on the limits of parental power.

The first is *Union Pacific Railway Company* v *Botsford,* an 1891 federal appeals court decision. The proceedings had their origins in a train journey in which one Clara L. Botsford occupied a lower sleeping berth. The upper berth collapsed and fell on her, causing extensive injuries. She sued the railway company and was given $10,000 damages. Three days before the trial, the company had applied for an order to compel Clara Botsford to submit to a surgical examination in order to provide information about her medical condition. She refused, and the federal circuit court in Indiana upheld her refusal.

The company appealed the decision, but the federal appeals court also upheld Clara Botsford, dismissed the company's application, and confirmed the results of the earlier proceedings, making in the process the following pronouncement: "No right is held more sacred, or is more carefully guarded by the common law, than the right of every individual to the possession and control of his own person, free from all restraint or interference of others, unless by clear and unquestionable authority of law."

Almost fifty-three years later, in January 1944, Mr. Justice Rutledge, in delivering the opinion of a federal appeals court on the bodily control of children by parents and guardians, used these words: "Parents may be free to become martyrs themselves. But it does not follow that they are free, in identical circumstances, to make martyrs of their children. . . ." The case, *Prince* v *Massachusetts,* contains a number of unusual features and has played an important part in the development of American law on transplants, though the facts of the case had nothing to do with medicine. The central characters were members of the Jehovah's Witnesses, a religious sect that has frequently had head-on collisions with the law over blood transfusions and surgical procedures. Indeed, the judgement of the court opened with the words: "The case brings for review another episode in the conflict between Jehovah's Witnesses and state authority."

Mrs. Sarah Prince of Brockton, Massachusetts, had been convicted of violating that state's child labour laws by arranging for her nine-year-old niece, who lived with her, to stand in the street and sell religious literature. Both Mrs. Prince and her niece, Betty Simmons, were Jehovah's Witnesses, and had followed the practice of regularly going into the street at night and distributing to passersby such magazines as *The Watchtower* and *Consolation.* As a result of doing this in the two weeks before Christmas 1941, Mrs. Prince was charged with two offences: furnishing Betty "with magazines, knowing she was to sell them unlawfully, that is, on the street," and "as Betty's custodian, permitting her to work contrary to law."

In the district court of Brockton, Mrs. Prince was found guilty and fined on both complaints. She then took the matter to the superior court of Plymouth County, where fines were reimposed, and after that to the Supreme Judicial Court of Massachusetts, where she lost the appeal. Finally she took her case to a federal court of appeals. In dismissing her appeal, and affirming the original convictions, the court not only made its celebrated comment about parents not having power to make martyrs of their children, but offered other observations on the limitations the law places upon parental power: "But the family itself is not beyond regulation in the public interest, as against a claim of religious liberty. . . . And neither rights of religion nor rights of parenthood are beyond limitation. . . . The right to practice religion freely does not include liberty to expose the community or the child to communicable disease or the latter to ill health or death."

The legal stage was now set for the enactment of ethical and moral dramas of the new transplant age. These began to appear with some frequency in the years following the first viable kidney transplants in 1954, when the medical profession showed an increasing ability to ensure high success with kidney grafts between living identical twins and between close blood relatives. The first reported cases in which judges were asked to order or permit the removal of kidneys from minors occurred in 1957 in Massachusetts, but it was another twelve years before a superior court was called upon to provide an authoritative statement of law and at the same time a just solution to a tragic family dilemma that came complete with many of the human and legal conundrums posed by live transplantation.

Strunk v *Strunk* entered the United States legal firmament via the Kentucky Court of Appeals on September 26, 1969, and has been under continuous discussion since that day. The court comprised seven judges, and the decision was made by a 4–3 majority.

Tommy Strunk was twenty-eight years old, married, and an employee of the Pennsylvania Railroad. He was also a part-time student at the University of Cincinnati. He and his brother Jerry were the two children of Mr. and Mrs. Arthur Strunk, who lived in Kentucky. Jerry Strunk was twenty-seven years old and a mental incompetent who had been committed to a state institution for the feebleminded, where there was both a school and a hospital. His mental age was approximately six years, and he was further handicapped by a speech defect which, according to evidence, made it difficult for him to communicate with people who did not know him well. Jerry was greatly dependent upon his brother Tommy, both emotionally and psychologically, to such a degree that a psychiatrist who knew him was of the opinion that Tommy's death would have had "an extremely traumatic effect upon him."

The Kentucky State Department of Mental Health expressed the opinion that it is necessary for every mental defective to establish a firm sense of identity with other persons. In the present case, Jerry Strunk identified strongly with his brother Tommy, according to departmental reports, and regarded Tommy as his model and his tie with his family. Jerry continually inquired when Tommy was coming to see him.

Tommy became ill with glomerulonephritis, a fatal kidney disease. His treatment involved being on a dialysis machine, but the time came in 1969 when this could not continue for much longer, and Tommy's only hope of survival was a kidney transplant. His doctors first gave thought to the possibility of using a kidney from a cadaver, but they had no way of estimating when, or even if, one might become available to them. They then raised with Mr. and Mrs. Strunk the subject of a live donation, as a result of which both parents and a number of their collateral relatives voluntarily underwent tests. None of them had tissue compatibility with Tommy sufficient to justify an attempted transplant.

As a last resort the family decided to test Jerry, and contrary to everybody's expectations, he turned out to have high tissue compatibility with his brother and to be eminently suitable as a prospective donor. This raised an immediate problem, which was how to go about removing a kidney from Jerry, who was obviously incapable of understanding what was going on, as well as legally incapable of giving an effective consent. Needless to say, the parents had decided that one of Jerry's kidneys should be given to his brother. The procedural problems were solved by first officially appointing a separate guardian for Jerry solely for the purpose of legal proceedings. Mrs. Strunk then petitioned a local court for an order authorizing the removal of Jerry's kidney. She said, in effect, that as Jerry was a ward of the state, neither she nor her husband had the parental power that might normally be expected. Her petition enabled the proceedings to be cast in an adversary form, so that both sides of the case could be put to the court. Jerry, through his guardian *ad litem* (court guardian), was able to put forward all available opposing arguments. In addition, the State Department of Mental Health sought and was given permission to join the proceedings as *amicus curiae* (friend of the court) in order to provide information and assistance.

The petition was first heard by a county court, which decided that the operation to remove Jerry's kidney was necessary and should take place. Doubts then arose about that court's power to make such an order, and two appeals ensued, the first to the circuit court of Franklin County, and the second to the Kentucky Court of Appeals. The appeals court had to decide whether legal power existed to authorize such surgery, and if so, whether the county court had been right to authorize it in this case.

In reviewing the evidence, the appellate court noted that departmental reports said that Jerry was aware he was able to help Tommy to continue to visit him, and that it was important to prevent him from developing guilt feelings if Tommy should die. The court's attention was also drawn to the opinion that Tommy was necessary to Jerry's life and was the only person who would be able to communicate with him after their parents' deaths. Other significant evidence was that Tommy did not have the capacity to survive several cadaver transplants, if there should be failures, and all members of the immediate family, plus the Department of Mental Health, recommended that Jerry should be the donor. Medical evidence was given that the risk of harm to Jerry from the surgical operation was about .05 percent, and the long-term risk to him of developing trouble with his remaining kidney about .07 percent. Throughout the proceedings, Jerry's guardian *ad litem* continually questioned the power of the state via the courts to allow the removal of an organ from the body of an incompetent ward of the state.

In giving judgement, the majority of the judges emphasized that the question before them was unique and had never previously been put to the highest state court or a federal court. They decided they possessed "inherent" power to deal effectively with the affairs of incompetents— inherited by the state courts from the High Court of Chancery in England, which in turn had received it by "delegation to the Chancellor of the Crown's right as *parens patriae* [guardian of the nation] to interfere in particular cases for the benefit of such persons as are incapable of protecting themselves." The power and its exercise, known in the United States as the "doctrine of substituted judgement," had been recognized there since 1844.

The doctrine of substituted judgement enabled a court of equity to authorize dealings with the property of an incompetent, on the assumption that the incompetent would choose to do the same thing if he were competent. To justify transactions of this kind, the court relied mainly upon the authority of *Re Earl of Carysfort,* an English case of 1840 concerning the property of a lunatic earl. A servant of the earl had been obliged to retire "by reason of age and infirmity." He did not have the means to support himself, and the earl was incapable of helping him. On an application to the Lord Chancellor's court, an order was made for the provision of an annuity out of the income of "the estate of the lunatic earl as a retiring pension to the latter's aged personal servant. . . ." The court was satisfied that the earl would have approved the gift if he had been capable of acting himself.

The Kentucky Court of Appeals was of the view that the doctrine of substituted judgement could be properly extended from the property of an incompetent to his body (thus equating the human body with property) and that a legal basis existed for making the kind of order sought in this case. After reviewing both the law and the facts, four judges decided that the removal of one of Jerry's kidneys for transplant to his brother was in Jerry's best interests, and should be allowed to proceed.*

Three of the seven judges dissented, and the eloquent words of Judge Steinfeld, one of the minority, perfectly illustrate the harsh ethical and moral quandary that faces anyone who honestly attempts to resolve a predicament such as that of *Strunk* v *Strunk:*

> Apparently because of my indelible recollection of a government which, to the everlasting shame of its citizens, embarked on a program of genocide and experimentation with human bodies I have been more troubled in reaching a decision in this case than in any other. My sympathies and emotions are torn between a compassion to aid an ailing young man and a duty to fully protect unfortunate members of society.
>
> I am unwilling to hold that the gates should be open to permit the removal of an organ from an incompetent for transplant, at least until such time as it is conclusively demonstrated that it will be of significant benefit to the incompetent. The evidence here does not rise to that pinnacle. To hold that committees, guardians or courts have such awesome power even

* Unfortunately the published law reports do not disclose the ultimate result of the surgery. In April 1980 the author was fortunate to speak to Mr. Morris E. Burton of Frankfort, Kentucky, who was attorney for Jerry Strunk in these proceedings. Mr. Burton was able to confirm that the transplant surgery duly took place and was successful, with neither party suffering ill effects. He said that he had had no direct communication with the Strunk family for a long time but he had seen two remarkable newspaper reports in the recent past which he believed to apply respectively to Jerry Strunk and Tommy Strunk. About one and a half years earlier he had read a newspaper report that Jerry Strunk had been killed in an institution for the retarded by a fellow inmate with a "toy" baseball bat. In the fall of 1979 he had seen an advertisement in a local newspaper showing a photograph of Mr. Tommy Strunk appealing for funds during "kidney week," and describing him as president of a local kidney association.

in the persuasive case before us, could establish legal precedent, the dire result of which we cannot fathom. Regretfully, I must say no.

In March 1972, the Superior Court of Connecticut was given the opportunity to apply the principles of both the Bonner case and *Strunk* v *Strunk,* and it willingly did so. *Hart* v *Brown* was a lawsuit brought by Mr. and Mrs. Peter Hart in relation to their identical-twin daughters Kathleen and Margaret. The children were seven years old, and only four months earlier Kathleen had developed a disorder called Haemolytic Uremic Syndrome, a form of kidney disease prevalent in young children. She was put on a dialysis machine in December 1971. By February 1972, she had developed a malignant type of blood pressure and malignant hypertension, as a result of which both her kidneys had to be promptly removed. She was then permanently dependent on a dialysis machine and her chances of survival for five years were no higher than fifty-fifty—unless she received a successful kidney transplant.

Both of Kathleen's parents would have been happy to give her one of their kidneys, but the medical evidence showed that the prospects for her survival with a parent's kidney were 70 percent for one year, 65 percent for two years, 50 to 55 percent for five years, and 37 percent for seven years. However, the prospects of success of a transplant from her identical twin were virtually 100 percent, on the basis of the records of the Human Renal Transplant Registry. The court was also much influenced by the descriptions it was given of the possible consequences of the immunosuppressive drugs that Kathleen would need to take in the event of a parental transplant—ranging from growth retardation and loss of the ability to walk, to hairiness and eye cataracts.

Not only was it Mr. and Mrs. Hart's wish that a kidney be taken from Margaret and transplanted to Kathleen, but approval was expressed in court by a clergyman, the two guardians *ad litem* of the twins, a psychiatrist, the physicians, and Margaret herself.

The court found that a transplant operation was necessary if Kathleen was to remain alive, that the risks, if the donor was Margaret, "were negligible" for both girls, and that such a donation carried an excellent prospect of good health and long life for both children. It also found that to subject Kathleen to a parental donation would have been "cruel and inhuman" because of the side effects of the immunosuppressive drugs. Drawing on the authority of *Bonner* v *Moran,* the judge declared that a nontherapeutic operation on a minor could be lawful if the parents consented. He also relied upon *Strunk* v *Strunk* and a series of Massachusetts cases in which decisions had been heavily influenced by the "grave emotional impact" upon minor donors of refusal to allow them to help their siblings. The removal of one of Margaret's kidneys for transplant to Kathleen was authorized.

The visibility of these legal landmarks was sufficient to influence the Superior Court of Georgia to follow the same direction in November 1973. *Howard* v *Fulton-DeKalb Hospital Authority* was a case in which the sick mother of a mentally retarded fifteen-year-old was advised to seek a transplant as the best way to cope with her own chronic renal disease. The daughter was the only relative who expressed a desire to donate a kidney, but both her age and her mental condition ("moderately mentally retarded") rendered her consent ineffective. The mother was willing for the transplant to take place; evidence was presented that the daughter was strongly dependent on her mother and would suffer severe physical deprivation and emotional shock if her mother should die because she was not allowed to give a kidney. The judge observed that a unique feature of this case was the fact that the consenting mother was also the prospective recipient; in these circumstances the principle that should be applied was not that of *Bonner* v *Moran* but of *Strunk* v *Strunk.* Under the "substituted judgement" doctrine, the court itself had power to authorize the removal of the daughter's kidney. With both mother and daughter willing, the court permitted the transplant to take place.

Yet not all American courts have received the message of *Strunk* v *Strunk* with approval. Indeed, some have rejected it outright. In both Louisiana and Wisconsin, superior courts have declined to follow it, the Louisiana Court of Appeals in November 1973 and the Supreme Court of Wisconsin in March 1975.

The Louisiana case, entitled *In re Richardson,* revolved around a seventeen-year-old mentally retarded boy with a mental age of between three and four years. He was otherwise quite healthy, but the average life span of persons with his condition was twenty-five years.

Roy Richardson lived at home with his parents and his sister Beverly, who suffered from kidney disease. Mr. and Mr. Richardson had several children, Roy being the youngest. By 1973, Beverly's kidneys had almost entirely ceased to function and her only prospect of living for more than a few months was a dialysis machine or a kidney transplant. Her doctors advised that a transplant was far preferable.

The Richardsons decided as a family that thirty-two-year-old Beverly should have a transplant if possible, and with one exception they all submitted to tissue-compatibility tests to determine the most suitable candidate to serve as donor. Roy turned out to be the one. Physicians who performed the tests estimated that the chances of Beverly's body rejecting one of Roy's kidneys within three to five years were no more than 5 percent, while with the other members of the family who had been tested the chances were 20 to 30 percent. They advised the court that the prospects of failure of a kidney taken from a dead donor "would be far greater than if one of Roy's kidneys is used."

Mr. and Mrs. Richardson brought legal proceedings before the civil district court for the parish of Orleans for an order authorizing Beverly's medical advisers to carry out the necessary surgery to remove one of Roy's kidneys and transplant it to Beverly. Both parents indicated their consent, as did Beverly. They even put Roy in the witness box briefly to answer questions by counsel and the judge. Medical evidence was given on success rates with transplants of kidneys, the prospects of additional transplants if one should fail, and the merits of transplantation as against dialysis. However, the judge decided there was insufficient evidence to warrant authorizing the request. He was much influenced by the medical evidence, which indicated that if one kidney transplant failed, it would be feasible to attempt others, and that Beverly's life did not depend upon obtaining one of Roy's kidneys or even a kidney from another family member, but could be sustained by a dialysis machine.

Mr. Richardson appealed to the Louisiana Court of Appeals, which heard further argument from both sides, and on November 6, 1973, rejected the appeal, affirming the lower court's judgement. The court of appeals spoke of the Kentucky decision in *Strunk* v *Strunk,* and indicated that it was prepared to agree that Louisiana's superior courts also possessed the same previously unused inherent power which the

Kentucky Court of Appeals had dusted up and produced from its legal armoury in 1969. However, both the facts of Roy Richardson's case and the Louisiana laws governing the administration of the property of a minor differed from those applied in Kentucky to Jerry Strunk. In Louisiana, the law was strongly protective of a minor's property and contained stringent rules preventing guardians and parents from dealing with it. "Since our law affords this unqualified protection against intrusion into a comparatively mere property right, it is inconceivable to us that it affords less protection to a minor's right to be free in his person from bodily intrusion to the extent of loss of an organ unless such loss be in the best interests of the minor," said the court. It then went on to find that the loss of a kidney in this case was not only not in Roy Richardson's best interest, but clearly against it. In those circumstances, neither Roy's parents nor any court could authorize a surgical intrusion into his body for the purpose of transplanting one of his kidneys to his sister Beverly.

Judge Gulotta added some remarks of his own on the principles that should be used to assess "best interests":

> I am of the opinion that before the court might exercise its *awesome* authority in such an instance and before it considers the question of the best interests of the child, certain requirements must be met. I am of the opinion that it must be clearly established that the surgical intrusion is urgent, that there are no reasonable alternatives, and that the contingencies are minimal. The requirements or prerequisites are not met in this case.

The Supreme Court of Wisconsin gave the principle of *Strunk* v *Strunk* much shorter shrift when its turn came on March 4, 1975. The case was *Lausier* v *Pescinski* and the facts were unusually tragic. The central character was also a mental incompetent, a thirty-nine-year-old chronic catatonic schizophrenic with a mental age of twelve years. Richard Pescinski had been an inmate of a state mental hospital since 1958, and although he was in contact with his environment, his behaviour, according to the medical evidence, showed "marked indifference." His mental disease was a flight from reality. The sick patient was Richard's sister Elaine, age thirty-eight and the mother of six children, all minors. A long-term sufferer from kidney disease, Elaine had both her kidneys surgically removed in 1970, and was then put on a dialysis machine in order to stay alive. By 1974, Elaine's condition had deteriorated, and it was decided that she must have a transplant. The medical specialist in charge of the case first gave consideration to her parents, Mr. and Mrs. Pescinski, as potential donors, but rejected them both because he was not prepared "as a matter of principle" to remove a kidney from a donor older than sixty. He also refused to consider Elaine's six young children because of their ages and his own "moral conviction." Elaine's sister Janice was a diabetic and therefore not eligible, and her brother Ralph, age forty-three, felt that his own family duties and health prohibited him from offering a kidney. He was a dairy farmer with ten children, nine of whom still lived at home; he had suffered a rupture of his left side and had a stomach disorder.

By this process, family attention focused on Richard. Permission was obtained to test him for tissue compatibility, and he was found to be suitable. Proceedings were then brought by his sister Janice Lausier, whereby the courts were petitioned, in a guardianship suit, to permit the removal of one of Richard's kidneys for transplant to Elaine. In the county court of Washington County, the judge refused to make this order, and Mrs. Lausier appealed to the Supreme Court of Wisconsin. The guardian *ad litem* appointed to represent Richard's interests refused at all stages to give a consent, thus ensuring that the question would be fairly and squarely faced by the court without biased assistance from Richard's side.

Chief Justice Wilkie, who delivered the court's judgement, flatly said there was "absolutely no evidence" that the interests of Richard Pescinski would be served by the transplant. That finding was not in itself decisive, for the court had also to consider whether it had the power to make an order of the kind asked for. This called for an examination of *Strunk* v *Strunk*, the only authoritative precedent capable of offering assistance. The chief justice was unimpressed by the judicial creativity of the Kentucky court in deciding that the doctrine of substituted judgement could be extended from allowing gifts of an incompetent's property to allowing gifts of his body contents. Unhesitatingly he said, "We decline to adopt the concept of 'substituted judgement' which was specifically approved by the Kentucky Court of Appeals in *Strunk* v *Strunk*. . . . We conclude that the doctrine should not be adopted in this state." Once the court decided it was not prepared to make a "substituted judgement," the ruling was obvious because no statute or legislative rule in Wisconsin spoke on this subject. The court was therefore unable to make the order, even if it had wished to: ". . . we further decide that there is no such power in this court."

The assembly by American law courts of rational legal principles regulating the removal of body parts from minors and legal incompetents may, in some respects, be likened to the construction of a building. The framework that gives it strength has been put together by the seminal decisions of the superior courts, such as those of Kentucky, Connecticut, Georgia, and the District of Columbia. However, the site preparation, windows, furnishings, finishing touches, and general appearance of the edifice have been provided by the courts "of first instance" of Massachusetts, a state that seems to be unusually well-endowed with vitality, intellectual energy, universities, and medical institutions. As early as 1957, twelve years before *Strunk* v *Strunk,* the first American court cases involving the use of minors as a source of tissue for transplant occurred in Massachusetts. There were three in that year, and each application was made to a single judge of the state's Supreme Judicial Court for authority to remove a kidney from a living minor for transplant to a twin. The surgery was to be performed at the Peter Bent Brigham Hospital in Boston, where the world's first successful kidney transplant between twins had taken place in 1954. The three applications were granted, and have been followed down to the present time by a continuous line of successful petitions to the same court and also to inferior courts (such as county probate courts), pursuant to which dozens of minors and other legally incompetent donors have been subjected to removal of kidneys, skin, and bone marrow. In one academic article published in 1975, a nonexhaustive list of nineteen authorizations by Massachusetts courts between 1970 and 1974 disclosed that kidneys were taken from seven incompetent donors and bone marrow from twelve. One Boston law firm filed thirty-one applications for bone marrow removal from minors between October 1972 and June 1978 with the Suffolk County Court and the Supreme Judicial Court. Similar procedures have been followed in Virginia and Maryland.

It is thus possible to demonstrate that the American judiciary since

1941 has created a unique and self-contained legal process, which appears to be without parallel in any other country, under which the law in a number of states will recognize and enforce claims made by the sick for the tissues and organs of living persons. As we have seen, agreement on the acceptability of such claims is not unanimous, nor are these laws accepted in every state. There are legislative restrictions too, such as a Michigan Act of 1970 that prohibits children under fourteen years of age from being living transplant donors; even when fourteen and over, the tissue may only be given to a member of the donor's immediate family. Nevertheless, the new judge-made principles are both widespread and firmly established in America. They were called into existence because the medical profession developed transplant techniques that could save desperately ill and dying patients, and because of the remarkable success rates possible in transplants between close family members with high tissue compatibility.

We have already seen that the law has never produced a comprehensive set of rules tailored to regulate modern surgery, but has extrapolated from the principles of assault and battery and other ancient canons. As a result, a doctor must obtain a consent to perform surgery if he is to have a defence to an action for battery. With minors and other incompetents, the question is how to obtain an effective consent even when the surgery is for their benefit. Difficulties multiply if the surgery is clearly not for their benefit. The old legal principles had been stretched to allow a person to consent to surgical assault and battery if it was beneficial to him, but they reached breaking point when the assault and battery would physically deplete him, even though his sole motive may have been to save another's life. When the person to undergo the bodily invasion was classified by society as one entirely lacking the capacity to look after his own affairs, the time had come to seek help.

In the face of laws like this, the American medical profession decided to pass the buck to the courts. It was too risky to decide without judicial backup that a consent on somebody's behalf was a real consent and on that basis to commence a surgical process that would inflict bodily harm on the subject while benefiting a third party.

After forty years of litigation, the majority of which has occurred since 1957, it is now clear that judicial backup can be obtained, but there is no unanimity on what is the true rationale of the decisions. The first step is usually to ask why the courts should have any power at all to make decisions which affect the bodily integrity of children. Why are parents not the sole arbiters of the welfare of their own offspring? The answer is that, generally speaking, parents do have this power, but not without exception. In *Bonner* v *Moran,* the court did not define the boundaries of parental power when the surgery is not for the benefit of the child—this point was not even discussed in the judgement—but there was nothing to suggest that John Bonner could have benefited in any way from the surgery.

The dilemma of parents becomes even more acute, and a severe conflict of interest arises, when two of their children are involved. How can a just decision be made by any parent whose child faces death or permanent illness but could be saved by the implant of an organ taken from a brother or sister? At this stage, exceptions to the general rule of parental dominion must be considered. In America, the state, acting through the courts, has traditionally intervened under its *parens patriae* power in the parent-child relationship when parents are not acting in a child's best interests or where they are unable to determine what those best interests are. The American medical profession has a noticeable habit of calling in the courts to exercise this power, under which the state is the ultimate guardian of every child.

Problems of this kind are common to many countries, whether the court authority derives directly from statutes, from traditional principles, or from both. There have been frequent cases in which parents have refused medical treatment for children on religious grounds: doctors and others have requested United States courts to order non-lifesaving operations, such as vaccination and removal of tonsils, as well as lifesaving measures, such as blood transfusion; and a variety of legal devices has been used to accommodate these requests, including temporarily depriving parents of custody, as well as a straight overriding of parental wishes. Yet there are numerous cases in the law reports in which parental desires have been allowed to prevail over considered medical advice. These include refusals of treatment, because the safety of the treatment and the certainty of benefit were not so clear, for rickets, deformed limbs, spinal fusion, and speech defects.

Obviously, the entire subject of medical treatment of minors is one of deep concern. As a separate matter altogether, there has been a general tendency to lower the age of legal capacity from twenty-one to eighteen. Clearly the purpose in fixing such an age is essentially to protect children, but all general rules can lead to hard cases and injustice. There are fifteen-year-olds, or twelve- or ten-year-olds, who are capable of reaching sound and sensible conclusions on difficult problems, while large numbers of their contemporaries would lack that capacity by any standard of judgement.

In England, it has been lawful since 1969 for physicians, surgeons, and dentists to accept sixteen-year-olds as patients without parental consent. In Australia, recent legislation provides for consent by minors of fourteen years to medical treatment, and since 1918, a statute on venereal disease has recognized two subcategories of "minor"—namely, "young person" and "child" ("young person" being between sixteen and eighteen years, and a "child" under sixteen). On the other hand, a court in British Columbia chastised a physician in 1970 for placing an intrauterine contraceptive device in a fifteen-year-old girl without first speaking to her parents. Further, when adopting Canada's Uniform Human Tissue Gift Act in 1972, the British Columbia legislature prohibited live-organ donation by persons under nineteen years, the statutory legal age in that province. The 1970 South African transplant statute permitted a minor over fourteen to give regenerative tissues without further consent. We have already seen that the state of Michigan also chose the age of fourteen as a minimum for transplant donors. In Norway, Sweden, and Denmark, legislation envisages live donation by minors only when there is "special reason," upon which a statutory tribunal adjudicates.

The law of minors exhibits a great deal of confusion. Most lawyers, and the public too, believe that under the common law of America and England minors are incapable of consenting to medical procedures unless authorized by special statutory power. But in the course of a penetrating law journal analysis in 1973, Dr. Peter Skegg of Oxford University found it necessary to say (and to prove) that "the view that, at common law, all minors are incapable of consenting to surgical operations, results from a fundamental misconception. . . ." On the other hand, in a paper on transplantation presented to the Royal Society of Medicine in 1969, one of England's Privy Council Law lords expressed the firm opinion that "the consent of a minor to act as donor ought not to be accepted. Nor can a parent or guardian lawfully consent to his child so acting." The speaker said he was aware of the thrust of American case law.

A fair comment is that whatever may be the correct law relating to minors, it ought to be simply and plainly put to the public. Decision-making on the medical treatment of minors, both beneficial and non-beneficial, and the use of a minor's body as a source of tissue and organs, not only is fraught with difficult problems in its own right, but is the subject of intense social concern.

It is fortunate that the United States courts have provided procedures to resolve hard family decisions. If we assume that the courts are fitting umpires for family quandaries concerning minors when the parents are unsuitable because of conflicts of interest or other reasons, it is reasonable to extend this fitness to other, comparable family dilemmas. What should be done about children or family members whose incapacity is due to mental and not chronological factors? If a person is both a minor and mentally incompetent, it seems that the court should play the same role as it would for a minor. If the person is of mature years and mentally incompetent, as was the case in *In re Richardson,* what is the role of the parents, or brothers and sisters? In such cases, the need for court intervention is in one sense even more pressing: These patients are frequently wards of the state, or of the court, and their affairs are entirely in the hands of the community. The court should therefore be involved in every major decision affecting their property or their bodies. If the incompetent person has not been committed to an institution or made a ward of the state, but lives at home with parents or family, the situation is comparable to that of a minor living at home, and turning to a court would have much to do with the question of whether medical treatment was beneficial or non-beneficial.

Assuming that application for authority to remove a body part from a minor or mental incompetent was made to an American court in one of the states that entertains such applications, what principles would the court apply today? Legal analysts have detected a lack of consistency in the judicial approach, and have suggested that the courts waver among three distinct principles, or tests. The first is the "best interests" test, which proceeds out of the "doctrine of substituted judgement." Under this test, the court's decision should be dictated solely by its perception of the best interests of the donor, and not those of somebody else, nor by sympathy for the proposed recipient or the family.

The second test, used in one or two Massachusetts cases, is that the donor's own consent "should be the sole criterion" in cases involving minors. As this test seems usable only when a minor is fully mature, it does not have much potential for general use. The leading case concerned a seventeen-year-old donor, and the court allowed his decision to donate bone marrow to be the sole criterion of its own authorization. Some writers would go further and argue that the entire concept of legal minority should be abolished: problems involving young people would then be dealt with as and when they arise, in the light of the facts of each case. One medical expert, writing in the *New England Journal of Medicine* in 1977, suggested that children of tender years were capable of participating in hard decisions in relation to their own bodies and used the expression "the traditional 'age of reason'—seven years": "One could show homage to the principle of autonomy by respecting the right of refusal as is customarily done with adults. . . . The traditional 'age of reason'—seven years—might be a reasonable lower limit."

In 1969, two eminent American contributors to the *Journal of the American Medical Association* made what was described by a speaker at the Third World Congress on Medical Law in Belgium in 1973 as "the incredible suggestion" that the age of seven would not be too young for a decision to be made on organ donation to an immediate family member, so long as the child was "found clearly to understand the nature of the procedure and the loss and risks" involved.

The third test American courts have used in these difficult cases has been called the "review of parental decision" test. This really means that the judge accepts in principle the importance of the parents' position, and tries to put himself in their place. It says not that the judge's primary duty is to the donor but rather that he should weigh the entire family dilemma and decide on the balance of family interests. It would thus be quite proper to sacrifice the interests of the prospective donor if they were outweighed by the potential benefits to the recipient. The leading case on this test involves an application to the Supreme Judicial Court of Massachusetts by the parents of two young children for authorization of a bone marrow transplant from their healthy six-year-old daughter to their ten-year-old son, who was suffering from aplastic anaemia. The daughter was obviously too young to have an informed opinion of her own. Disregarding the doctrine of "substituted judgement" and the "best interests" test, the court determined that parents have the "primary right and responsibility for deciding the delicate question." The judge regarded his role as no more than one of review of the parents' decision in order to ensure that it was not unreasonable. Finding that it was a decision that they could fairly come to as a family matter, he approved the bone marrow transplant.

Whether or not a judge consciously decides to use any particular test, he might in many cases be unable to ignore the plight of the family and the suffering of the sick, potential recipient. In fact, the impression one gets from these American reports is that the judgements are characterized by humanity and compassion rather than blind adherence to one principle or test. And legal proceedings are generally not begun until all other avenues are blocked and a family has reached the desperation point. It is unlikely that any family with a seriously ill or dying member would go to court in a lighthearted or superficial vein.

Yet, as harrowing as these cases may be for everybody concerned, the judges and the lawyers who have created this extraordinary body of law must be ready themselves to be judged. Possibly the very intensity of the tragedies and the plight of the families have prevented judges from seeing the "wood" and have made them concentrate unduly on the "trees." Time and time again, in analysis of these decisions, commentators have made observations like the following:

> The irony implicit in any standard that imposes such an obligation upon minors and other incompetents, but not upon adults, is that its use would take advantage of a judicial proceeding designed to protect the most vulnerable in our society in order to exploit them in a way that adults cannot be exploited.

No adult can be forced to donate a kidney. How, then, can it ever be justifiable to oblige a child to do so? It is very difficult to answer this criticism. Mr. Justice Rutledge's denial to parents of the power "to make martyrs of their children" should not be forgotten.

Some critics flatly deny there can ever be "benefit" to a person in losing an organ. This criticism is especially scornful of the practice of producing psychiatric evidence of potential "psychological benefit" to the incompetent in giving a particular body part to a sick family member; such evidence has been described as "the unpalatable charade of a parade of psychiatric experts finding a 'benefit' in what is patently nonbeneficial to the donor." Another powerful criticism emphasizes the purely theoretical nature of all consents given on behalf of children, and insists that nobody should ever be permitted to deprive a child of a vital organ without his intelligent comprehension.

Again, many doctors have drawn attention to the "coercive forces" generated within a family. Somehow or other, they say, in cases where there is more than one eligible donor, the family tends to fix upon one member as the "victim" and to exert tremendous mental and emotional pressure on that person. Another criticism claims that what often happens inside a family, and outside, is a balancing of social worth, whether consciously or not. This can have dangerous results for a mental incompetent who is eligible as a donor to a sick brother or sister, especially if the latter has heavy family responsibilities. Many people believe that tissue removal from such donors should be totally prohibited. In one leading case involving a feebleminded donor, his own mother told the court that she believed that at last his life had been given a purpose, and this was to serve as an organ donor to save his brother's life. The consequences of the deliberate application of this principle throughout society would be horrifying. In this particular case, the court did order that an organ be taken from the incompetent, but for other reasons. Some who oppose the American "incompetent removal" cases conclude that the only proper course is total prohibition, for true justice can be done only by legislation that bans without exception all live donation by incompetents and anybody subject to coercive pressure from others, such as prisoners. This extreme position holds that the prohibition should extend to all live donation within a family, whether by a minor or not. With intrafamily donations legally impossible, it is suggested, all guilt feelings and trauma would be automatically eliminated. Society would carry the responsibility. At the same time, the donation of body parts by live, consenting, unrelated adults would be lawful, being a different thing altogether.

Before leaving the subject of body-part removal from minors and other legal incompetents, it would be useful to glance at prisoners and people in comparable positions of disadvantage as sources of tissue for therapy. There is as yet no sign of property claims or widespread public pressure upon prison inmates for their bodies or body contents, at least not in the West; but they are peculiarly vulnerable. Documented accounts of medical experiments and similar practices with prisoners, and of abuses of the aged and sick in public institutions, serve to warn of the ease with which people may be dealt with as objects owned by society.

It is arguable that the "dehumanization" of certain classes of people by others is one of the greatest dangers to the individual in the modern world. There is ample evidence of the classification by governments, revolutionaries, guerrillas, and others of entire groups of human beings as unworthy of membership in society, and therefore expendable by any means, whether deliberate neglect, slave labour, torture, or destruction. These groups have comprised both small numbers, such as victims of terrorists and captives who are dissidents or who have otherwise failed to "conform," and large, such as ethnic communities, entire social classes, or even a national population, as in Cambodia. The overwhelming events that have followed this sort of behaviour bear little quantitative relationship to the performance of experiments upon selected inmates of a hospital for the aged, but the lesson is there to be learned from any treatment of a person as though he were simply a thing.

One academic investigator into the participation of prisoners in human experimentation reported in 1974 that half of the fifty states in the United States permitted drug research on convicted criminals. Detained men and women are attractive as subjects for experiment because of their continuous availability and the possibility of strict dietary and environmental control. These advantages, plus prisoner cooperation, led a large drug company to build a pharmacological laboratory and a ten-bed "metabolic ward" on the grounds of a state penitentiary in Michigan. Although prisoners cannot be legally compelled to participate in experiments or medical procedures, subtle economic pressures are often at work. For example, many prisons in the United States do not supply inmates with basic commodities such as toilet articles and books, and therefore they must earn money to buy these items.

A famous demonstration of prisoner participation in medical therapy occurred in the state prisons of Colorado in the 1960s, when a number of inmates gave kidneys for transplantation. No pressure was applied by the administration or the medical profession, as shown by the fact that some four thousand prisoners were approached and fewer than one hundred expressed interest. The only means used were notices placed on bulletin boards. Of the volunteers, only a fraction were actually accepted and underwent kidney removal. They were all tested psychologically to determine motivation, and it was noted that most of the volunteers were guilty of relatively minor crimes such as car stealing. Many of them had only a few weeks or months left to serve of their sentences. The signed consent form for the operations contained two clear stipulations: there was to be no pay, and no reduction of time served in prison. These conditions contrast strikingly with programs of an earlier era. The famous French physiologist Claude Bernard (1813–1878) revealed in his writings practices in France whereby condemned men were allowed and even encouraged to undergo dangerous surgery in exchange for full pardon. Even so, worries about the truly voluntary nature of the participation and the possibilities of abuse caused the organizers in Colorado to discontinue the program in 1966—but not without regrets. One of the doctors involved told an international symposium in the same year that the benefits had included "the procurement of considerable scientifically controlled data which can be brought to bear on future problems"—new information on the characteristics of transplants between nonrelatives as compared with intrafamily transplants, new information on immune rejection, and increased experience with preoperative tissue typing.

Society's readiness to sacrifice principle for expediency when it needs to obtain blood is also evident in relation to blood donation by prisoners, at least in the United States. Richard Titmuss, writing in 1970 about "the captive voluntary donor" of blood, showed that in

Britain a little under 1 percent of the national total of blood donations in 1967 came from 14,903 prisoners, who gave voluntarily and who earned no reprieves or rewards of any kind. In America, he had found evidence of increasing resort to prisoners for blood donation, and estimated that they contributed 2.2 percent of all blood given to the American Red Cross in 1967 (some prisoners were paid money, some gave voluntarily, some were given favourable references on their records, and some received other benefits). A number of states had laws that gave remission of sentences calculated according to the quantity of blood donated—for example, Massachusetts under a 1965 Act gave a remission of five days for every pint. Titmuss also offered evidence of considerable use of prisoners for testing pills and vaccinations, and reported a suggestion that between a quarter and a half of all first-phase drug tests were carried out on prisoners.

Other people commonly used as subjects in experimentation and medical programs include students, nurses, laboratory technicians, persons under discipline such as members of the armed forces and police, hospital and institutional patients. Many volunteer, responding to conscious or subconscious feelings that they must please a superior or live up to expectations, or even because of misplaced confidence. A heavy ethical and moral responsibility lies upon those who occupy the "superior" position in such programs. Personal freedom can be much more easily denied to individuals in out-of-balance relationships. This does not mean that experiments in which prisoners or military personnel participate are automatically objectionable. In November 1979, considerable publicity was given to a twenty-five-year secret program of experiments that had been conducted until 1975 by the U.S. Army, at a cost of some $80 million, testing drugs intended for chemical warfare. The program had secured the participation of hundreds of prisoners, 7,000 military personnel, and 1,074 civilians.

Some of the most notorious American examples of the abuse of the disadvantaged have concerned old people, mentally retarded children, and black citizens. One, which has been called the *Jewish Chronic Disease Hospital Case,* occurred in the 1960s in New York City, where twenty-two elderly patients in a hospital for the chronically ill were injected with cancer cells to determine how quickly rejection would occur in old people as compared with healthy younger subjects in a prison, on whom experiments had already been carried out. The patients were not told what they were being injected with, so there was no informed consent. In 1963 and 1964 successful charges of fraud and unprofessional conduct were brought against the responsible medical personnel.

In another case in New York State (1964 and 1965), mentally retarded children were given an extract made from fecal material which contained hepatitis virus, and in a subsequent experiment, others were directly injected with infectious hepatitis virus. Another experiment was a long-term study of syphilis conducted in Alabama over a forty-year period ending in 1972, which involved some six hundred black men, many of whom were deliberately deprived of treatment for their disease and some of whom were thought to have died as a result of this deprivation.

Despite these recurrent examples of physical abuse of disadvantaged persons, there seems to be no evidence that society today approves of subjecting prisoners, members of the armed forces, or others rated as second-class citizens to bodily risks in experiments in preference to normal citizens. Generally, a great deal of protection is available, from sound ethical principles to sound methods of official administration. But there is another side of the coin, and its message is that a blanket ban preventing all disadvantaged persons from playing their part in community activities simply because of the possible abuse of their weakness would be unjust. One philosopher put it thus: "I am one who happens to believe that prisoners have not been and should not be drummed out of the human race." On this hypothesis, the disadvantaged ought to be allowed to participate in appropriate programs, provided that their special interests are safeguarded.

The Council of Europe understood the issues, as a memorandum accompanying its 1978 model rules on body-part removal shows:

> A number of experts proposed not to admit imprisoned persons as donors at all, fearing that such donations might be given in expectation of a pardon or a good conduct report enabling them to secure an early parole. The majority ... however, preferred not to bring any exceptions for imprisoned donors. . . . If an imprisoned person gives his consent freely his wish must be respected; if, however, he is giving his consent under coercion

or in order to obtain a reward against it, then his consent not being given freely, his donation cannot be accepted and no removal can take place.

Close attention should be devoted to the particular vulnerabilities of the young, the helpless, and the socially handicapped when medical activities and experiments call for access to human bodies. Their need for protection shows how the unrestricted exercise by one person of a right, a freedom, or a liberty can easily lead to the denial of rights, freedom, and liberty to others.

II.

As we saw earlier, the latest laws governing how human tissues are obtained place greater importance on community requirements than on the individual and his autonomy. Much judge-made law can be interpreted as reflecting a similar philosophy.

For these reasons, the case described below could prove to be the historic watershed from which a community-oriented attitude to tissue donation could spread throughout the United States and across national boundaries. On the other hand, it could be taken as a rocklike declaration of the primacy of the principle of personal choice. Its facts alone give it a rare importance, because it is the first known instance of an adult, competent citizen asking a court to compel another adult, competent citizen to surrender part of the contents of his body. The fact that the applicant lost the case does not make it less notable. What is momentous is that the case could be brought at all, and that it "got off the ground" without resort to artifice or manipulation of forensic procedures. The plaintiff made his demand as a matter of direct legal principle, and it was received and heard by the judge on its merits.

Robert McFall of Pittsburgh, Pennsylvania, was overwhelmed by the symptoms of aplastic anaemia in June 1978. A nightmare began for the thirty-nine-year-old bachelor when he began to develop bruises after bumping into objects during his work installing insulation materials in confined spaces in buildings. The bruises would not go away, and soon he began to have nosebleeds that continued for hours at a time. He went to a local hospital in suburban Pittsburgh, where the doctors diagnosed aplastic anaemia, a rare, almost certainly fatal disease of the bone marrow and blood. The prospects of death after contracting this disease have been put by some medical studies at 90 percent, with an average survival period somewhere between three and four months. There is only one real source of cure, and that is a transplant of compatible bone marrow. This transplant gives a good expectation of complete recovery. Without it, the patient must expect to die.

The statistical likelihood of finding compatible bone marrow is almost one in sixty thousand. In practice, the prospect is hopeless, because no means exist for testing the community. There are as yet no computerized banks containing comprehensive national tissue information (though in some parts of the world, for example, at Westminster Hospital in London, computerized tissue banks are being built up and already contain information about thousands of prospective donors). On the other hand, the prospect of finding tissue compatibility inside a family is far higher, and increases with the closeness of the relationship.

Robert McFall had three brothers and three sisters. They had all gone their separate ways following their mother's death in 1949, and there had been little family communication after that time. By means of computer checks through driver's license records, they were all traced, and agreed to submit to tissue-typing tests. None of them turned out to be a compatible donor. It was then decided to enquire whether a first cousin of McFall, David Shimp, a crane operator in a steel mill, would agree to be tested. Shimp was aged forty-three and married. When both men were younger they had gone to camps together and had shared many experiences.

Shimp agreed to undergo a preliminary test but did not bother to tell his wife. The test proved to be positive, suggesting that Shimp's bone marrow would be a perfect match for Robert McFall. A second test was arranged, but Shimp cancelled the appointment. He had changed his mind, and from that time onward refused to have anything more to do with the affair. According to reports, Shimp said his wife was angry that he had taken the test without discussing it with her, and wanted him to discontinue his participation. His mother had expressed the same wish. One report said that Shimp had been influenced by a dream that if he went into hospital for the bone marrow removal, he would never come out. Friends and other relatives put great pressure on him to proceed with the tests, but he would not budge. It was even said that he considered bringing legal proceedings to stop harassment, because the story had gotten into the hands of the media, which gave it considerable publicity. However, it was Robert McFall who first resorted to the courts.

In the last week of July, McFall sued David Shimp in Allegheny County Court, Pennsylvania, asking for an order that would compel Shimp to submit to further tests, and eventually to the removal of a quantity of his bone marrow for transplant to McFall. Time was now all-important for McFall, and the normal delays of court hearings too risky. His lawyers asked for an urgent preliminary injunction, which, if granted, would direct Shimp forthwith to undergo the further tests. In this atmosphere events moved rapidly, and *McFall* v *Shimp* was dealt with and disposed of on July 25 and 26 by the Civil Division of the Allegheny County Court, Judge John P. Flaherty, Jr., presiding.

The plaintiff's brief was a document of originality and persuasion, skilfully prepared by his attorney, John W. Murtagh, Jr. Its opening words went straight to the heart of the matter, posing for determination an issue as profound as any that could be put to a court of law. The judge was asked, in so many words, to determine whether society may overrule a citizen's claim to an absolute right to his bodily security in order to save the life of one of its members. The brief submitted, for reasons it set out in detail, that the answer "is and must be 'yes.' " It then tackled some of the medical questions, asserting that the removal procedure was medically safe, would at most result in minor and temporary discomfort, and would deprive the defendant of nothing but his time because bone marrow is a regenerative tissue that promptly replaces itself.

McFall's lawyers had found no precedent or comparable case that could directly assist the court, and the judge himself later commented that "a diligent search has produced no authority." Accordingly, the claim for legal relief was put in fundamental terms, based on morality, ethics, custom, scholarly legal pronouncements, and judicial opinion. Of the decisions examined earlier in this chapter, the brief referred to only one, *In re Richardson,* the case in which the Louisiana Court of Appeals refused to sanction the removal of a kidney from the mentally incompetent seventeen-year-old Roy Richardson for transplant to his sister Beverly. It was argued for McFall that the facts in *Richardson* were quite distinguishable: Beverly Richardson was capable of staying alive by use of a dialysis machine, and her life did not depend on obtaining a transplanted kidney, let alone one from her brother; there was evidence that she could return to a dialysis machine and wait for another donor to become available; finally, the case had been about removal of an entire organ, not removal of regenerative tissue like bone marrow. As far as Robert McFall was concerned, it was all or nothing: either he was given bone marrow from the defendant or he would die.

McFall's case then cited some well-known circumstances in which bodily integrity is lawfully disregarded because of overriding social considerations: public health requirements for vaccination and quarantine; criminal law powers to take hair, blood, clothes, and semen; marriage law requirements of blood tests; defense law requirements of military service; and compulsory assistance to law enforcement officers in emergencies. To these McFall sought to add his own case as representing a new category. To demonstrate that the court could, by reference to principle and precedent, extend the law in this fashion if it wished, the plaintiff produced the fruits of some extremely original research.

Power to make an order of the kind requested was traced back some seven hundred years from the Allegheny County Court, through the Pennsylvanian and United States legal systems, to the ancient English Courts of Chancery and the powers to dispense justice granted to those courts in the reign of King Edward I. This English king ascended his throne in the year 1272, and in the thirteenth year of his reign, Parliament passed the statute now known as the second Statute of Westminster. It contained the following provision: "Whensoever from thenceforth a writ shall be found in the Chancery, and in a like case falling under the same right and requiring a like remedy, no precedent of a writ can be produced, the Clerks in Chancery shall agree in forming a new one; lest it happen for the future that the Court of our lord the king be deficient in doing justice to the suitors." The legal inheritance of the Commonwealth of Pennsylvania includes the powers delegated by the second Statute of Westminster. Writing in the *Harvard Law Review* in 1908, James Barr Ames described the statute as "a perennial fountain of justice to be drawn upon" wherever the absence of a remedy "would shock the moral sense of the community."

The question was whether Robert McFall's claim should be recognized by the courts, and whether the law should regard David Shimp as having a duty toward him. "Has the duty the Plaintiff seeks to impose upon the Defendant ever been recognized in law or equity?" asked the brief. In support of an affirmative answer, reliance was next placed upon the so-called Rescue Cases.

The legal principle of rescue recognizes the social duty of a citizen to act positively to attempt to rescue another who is in personal danger. A yachtsman may be found to have a positive duty to try to save a drowning man. American and British laws have not favoured the rescue concept and have been reluctant to equate moral with legal obligation. Generally speaking, their approach has been to recognize that certain relationships should produce legal duties and obligations, for example, the relationship of doctor and patient. They have been slow to build specific legal duties on the foundation of general moral concepts, particularly when this might result in conflict with "individualist" philosophy. In the words of one American judge, "common law courts have been reluctant to impose affirmative duties on individuals even in situations in which most people would feel under a moral obligation to act."

By contrast, European nations, with a civil law tradition stemming from ancient Rome, have found it much easier to accept the rescue principle. They have tended to avoid the particularity of Anglo-American statutory drafting, and prefer writing laws in broader, more conceptual terms. Socialist countries have also been inclined to accept the rescue principle, no doubt because of their habits of dealing with citizens in terms of social duties owed rather than personal rights possessed. Writers have noted the readiness of French law to impose a duty to rescue, and back it up with penal sanctions, particularly in the years since World War II. One commentator neatly described the French attitude as representing the philosophy "that the evil of indifference to another's danger and consequent failure to extinguish it is more serious than possible infringements on liberty which the requirement of action might entail." Similar legal attitudes exist in many countries including Denmark, Holland, Italy, Norway, Poland, Rumania, and the U.S.S.R.

McFall's lawyer claimed that in recent years American and English lawmakers had undergone some change of heart. Examples were provided of the "ebbing of the strongly individualist philosophy of the early common law," and of cases in which courts had countenanced exceptions to the general rule that refuses to impose a duty to rescue. On the subject of yachtsmen, he was able to point to a decision in which a court held that a yacht owner whose guest fell overboard was under a positive duty to rescue the guest. He put to the court that it was possible to detect a growing Anglo-American acceptance of the principle that legal consequences should attach to conduct that displays indifference to the peril of a stranger. On McFall's behalf, he also put forward and supported a model set of standards proposed in 1965 by a prominent advocate of the rescue principle. These standards, which did not reach the statute book, contained specific suggestions for the provision of medical aid by means of blood transfusion. The basic proposal was that a person should have a legal duty to attempt rescue whenever another was in imminent danger and the first person was the only practical source of help. The duty would apply only if the danger would lead to substantial harm to person or property, and the risk to the rescuer would be "disproportionately" less than the prospective harm. On the subject of blood transfusion, no objection was seen to a general rule that citizens should be placed under a community duty to give blood. The drafter urged that, at the very least, any blood donor could logically be placed under a duty to continue to give blood, for by giving his tissue in the first place, he indicated that his bodily security was "subordinated to some other interest"; it was accepted, however, that a person opposed to blood transfusion should not normally be held liable for failure to give blood even if it resulted in loss of life. When these standards were formulated in 1965, the safe removal of bone marrow had not appeared as a lifesaving procedure, but presumably the same philosophy of compulsory donation could be applied to bone marrow donation, and to any other body tissue or organ which, as medicine develops, may be removed without impairing a person's health or well-being. It should not be forgotten in considering this argument that right now a person with one healthy kidney is as acceptable to life insurance companies as a person with two.

The plaintiff's brief argued that Shimp's behaviour in undergoing the first test had placed him in the same position as the blood donor who has previously given blood. By permitting himself to be tissue-typed and by demonstrating a four-tissue match with his cousin, he had obligated himself to continue. The brief claimed that Shimp had "cruelly abandoned the Plaintiff after the Plaintiff was allowed to hope for a successful end to his ordeal," and should be compelled to continue to offer aid because the plaintiff had thereby been exposed to the risk of greater harm: McFall's chances of cure had been diminished due to the delays caused by Shimp's initial embarkation on a program of assistance and his later refusal to proceed.

The brief ended with the plea that the court, as the voice of society, should not in the name of the defendant's bodily security abandon Robert McFall to a short, medically dominated life and certain death. "Our noblest tradition as a free people and our common sense of decency, society and morality all point to the proper result in this case. We respectfully suggest that it is time our law did likewise."

On July 25, 1978, in a preliminary hearing, Judge Flaherty had to decide whether Robert McFall had disclosed any kind of legal case at all and whether David Shimp had an obligation even to offer a defence. He considered the matter and heard medical evidence of the plaintiff's low chance of survival, the "minimal risks" in bone marrow removal, and the fact that the plaintiff would have at least a 50 percent chance of cure after a transplant from the defendant. The judge then directed that Shimp's attorney file a brief setting out the reasons why he should not be ordered to give the bone marrow. A hearing was fixed for the next day. McFall had negotiated his first legal hurdle.

The essence of Shimp's defence was contained in his attorney's argument that the law of Pennsylvania did not impose upon him any duty to help his cousin. Whatever had been said about "minimal risks" of bone marrow donation, the fact remained that the risks existed, and it could be dangerous. Though it is regarded medically as safe, bone marrow removal involves general anaesthetic and extraction of the marrow from the pelvic bone by means of a specially designed needle, which may be inserted as many as two hundred times in order to obtain the required quantity. This process can have a strong psychological effect upon the donor, particularly if he has a fear of surgery, or a fear of losing part of his body, and can cause him to develop hostility toward the recipient.

Even if no risk existed, his client was still under no legal obligation to come to anybody's aid, said Shimp's attorney. Finally, he said, McFall's claim was suspect because it rested upon a view of what the law ought to be, not upon the reality of the law as it then was.

Judge Flaherty delivered his final opinion on July 26, 1978. He accepted the evidence that Shimp was the only suitable donor and that McFall was unlikely to survive without the transplant. He also agreed that the Allegheny County Court is a successor to the English Courts of Chancery and derives power from the second Statute of Westminster.

The judge analysed the American common law attitude to the rescue principle and noted that the common law had consistently adhered to the rule that one human being is under no legal compulsion to aid another who is in distress or danger. He said that "on the surface" the rule "appears to be revolting in a moral sense." However, he felt that mature reflection would demonstrate that it is of "the very essence of our free society," which takes respect for the individual and his protection as its first principle. The judge contrasted this philosophy with that of societies which hold that the individual exists to serve the community as a whole. On the other hand, in a free society like the United States, moral conflicts such as that raised by Robert McFall are bound to happen. The judge considered that the true decision in this case was a moral one, and rested upon the defendant Shimp, adding that in the view of the court his refusal was "morally indefensible." He was, however, not prepared to compel Shimp to submit to bodily intrusion. "To do so would defeat the sanctity of the individual and would impose a rule that would know no limits and one could not imagine where the line would be drawn.... Forceable extraction of living body tissue causes revulsion to the judicial mind. Such would raise the spectre of the Swastika and the Inquisition, reminiscent of the horrors this portends."

Robert McFall lost the case. About three weeks later he died.

It is a matter of interest that in McFall's case no link was made with the line of decisions on tissue removal from minors and mental incompetents, except for the reference to *In re Richardson.* Possibly this was because these decisions were seen as precedents for the courts representing the state to act as the ultimate guardian of children and the mentally retarded. To the extent that others, as well as the McFall court, have viewed them in this light, their significance for an assessment of the human body as property and its need for protection has been overlooked. Even if such a link had been made in *McFall* v *Shimp*, the court might have recoiled from the minor-incompetent cases and still found against the plaintiff.

McFall's case was certainly the first case of its kind between two competent adults, but it was not the first opportunity for a United States court to order tissue removal from a person who has given no consent. Although there is a clear difference between a strong refusal to donate, such as David Shimp's, and a failure to consent, such as Jerry Strunk's, the difference is only one of degree. Of course, the law pretends otherwise. The legal fiction is that the court's agreement on behalf of Jerry Strunk, the twenty-seven-year-old incompetent with a mental age of six, amounted to Jerry's own consent. Thus it could be claimed that *McFall* and *Strunk* are not comparable and that we are not "comparing apples with apples" because in one there was a positive refusal and in the other there was a consent.

Whatever distinctions may be drawn between these two apparently different cases, or between either of them and any of the others, the abiding (and perhaps inconvenient) lowest common denominator remains: they all have as their purpose the assertion of dominion over the body of a living person and the removal of some of its contents in circumstances where the person has given no consent of any kind or is incapable of giving a comprehending consent.

The decision in any particular case will, to some degree, depend upon the personal philosophy of the judge, because in this field there are no statutory rules and no settled judicial law. The judge in *McFall* v *Shimp*, John P. Flaherty, Jr., was appointed a justice of the Supreme Court of Pennsylvania not long after that case. He has handed down other judgements which contain the clearest evidence of a humane, "individualist" personal philosophy. In June 1977, in the case of Mrs. Susan Frye, he made a difficult decision, while still in the county court, not to force a lifesaving blood transfusion upon a dying young woman whose religious belief was that a transfusion would amount to "a sacrilege and violate her relationship with God." The judge visited Mrs. Frye in hospital on a number of occasions. After satisfying himself that she fully and intelligently understood that her refusal to accept transfused blood would lead to her death, he declined to order the administration of a transfusion because of his respect for her decision and his own belief in "the sanctity of the individual." "If the state is able to force one to have a blood transfusion when one of sound mind refuses, what else can the state do? This frightens me, and it should frighten all who read this. Where does it end?"

As clearly as the judge has declared his personal views in these decisions, it is equally clear that, being expressions of opinion, they would not necessarily be shared by all judges. But there are other notable aspects of *McFall* v *Shimp*; the facts that the proceeding itself was readily entertained by the court, that Robert McFall was accorded *locus standi* (standing to sue) and that the "cause of action" upon which it was based did not offend either the procedural or substantive requirements of Pennsylvania law. In other words, there is no reason why the submissions in the brief filed by McFall's attorney may not survive to be considered and accepted by another court on another day. Greater emphasis on the difference between regenerative and nonregenerative tissues might be decisive in producing an eventual breakthrough. There could be no more dramatic illustration of the pressures and arguments that might cause a reversal of the *McFall* principle than the outcome of Mrs. Frye's case. The application to Judge Flaherty for compulsory blood transfusion had been made by her husband. After the judge decided not to override Mrs. Frye's expressed wish and not to order the transfusion, the husband's attorney immediately approached an appeals judge, who overruled the decision and made the order. The transfusion was administered a day or two later. Mrs. Frye did not die.

The United States now possesses judge-made laws on the use of the living body as a tissue source that could rapidly move into conflict with statutes on the use of dead bodies. The Uniform Anatomical Gift Act, in force in every state, deals only with the dead and rests four square on consent, prohibiting the removal of any body material without it. For adults, the consent may be that of the deceased or of relatives. For dead minors and mental incompetents, the consent may be that of parents. Ironically, this jealously preserved requirement for consent may be responsible for much of the acute shortages of organs and other tissues. The constant question is how to increase supply, and one of the answers now being given with greater frequency is "compulsion," the antithesis of consent.

By failing to regulate donation by living persons, the Uniform Act has left lawmaking to the courts. We have now seen some of the things that have happened. It could be said that live donations are still governed by the same philosophy as that of the Uniform Act, and that the cases involving minors and incompetents depend upon the requirement of consent just as the Act does. They show that without a court's agreement, those lacking legal capacity should not be subjected to tissue removal, and with a court's agreement, they may be; in law, this adds up to consent. But is it really the same? Surely it is wrong to apply this kind of logic. On any human approach, there is a vast difference between taking a kidney from a child's cadaver with parental consent, and taking a kidney from a six-year-old child with parental, judicial, or any consent other than a comprehending consent of the child itself. The use of the same descriptive words does not alter the truth of the matter. It can be both inadequate and misleading to say simply that "consent is necessary for the removal of parts from all bodies of all persons of all ages both dead and alive." The factors that may make up a "consent" should be defined, and the enormous factual difference between allowing a family member to authorize tissue removal from the body of a dead relative on the one hand, and from a living relative on the other, should be appreciated.

The time has arrived when direct comprehending consent to allow tissue removal from certain classes of living donors is no longer necessary throughout the United States. Orders for the compulsory implant of tissue both to adults and minors are not infrequent, at least with blood. With *McFall* a further stage was reached in relation to living adults. An obvious comment upon *McFall* is that the unequivocal decision confirmed the necessity for consent to adult tissue donation. Yet, for reasons given above, *McFall* may not be the bulwark for personal autonomy that it appears to be at first glance. Rather, it is possible to suggest in relation to live tissue donation that United States law stands at a crossroads.

Because the adversary system encourages presentation of both sides of disputes, and court decrees do not have the rigidity that characterizes statute law, there is scope for variation in judicial decisions. American courts are now in a position to dispense with the general requirement of consent or, if they refuse to go to that extreme, to introduce on a limited basis principles of peremptory or compulsory donation.

There is also another legal principle they could call on. Let us consider an unlikely, hypothetical case. A popular President of the United States is on a camping vacation in a remote mountain area prior to a summit meeting which could be important for world peace. He sustains a serious injury in an accident and loses a great deal of blood. The President has an unusual blood type and there is only one person in the vicinity with matching blood. The President cannot be moved far and will die without a transfusion, but the man who can save him refuses to donate a drop. Should he be compelled to?

The possibility that the law would allow the unwilling blood donor to be physically overpowered and enough blood taken from him to save the President's life is by no means remote, even disregarding the cases we have been discussing. Many nations, including the United States, know the defence of "necessity," which is a relative of the "rescue" principle. This controversial concept permits "necessity" to be pleaded as a defence to criminal charges when a person commits a crime because it is necessary to do so in order to avoid a greater harm. The lawbooks are replete with extraordinary cases, both real and imaginary, that illustrate it. Two of the most remarkable occurred in the nineteenth century, one in America and one in England. In 1842, an American grand jury had to decide whether a seaman named Holmes had committed murder. Holmes' ship was wrecked in a collision with an Atlantic iceberg, and he managed to escape in an overcrowded lifeboat with thirty-two passengers, the first mate, and seven other seamen. The lifeboat would certainly have sunk with this number on board, so in order to save as many as possible, Holmes and some other seamen, on orders, threw overboard a number of male passengers. The boat, remaining passengers, and crew members, were rescued. Because of the circumstances, the grand jury refused to indict Holmes for murder although it did regard his action as manslaughter. He was given six months in prison.

The English case also involved a shipwreck—a yacht which in 1884 sank one thousand six hundred miles from the Cape of Good Hope. Three seamen and a cabin boy of seventeen were cast adrift in an open boat. Eight days after the food ran out and six days after all the water was consumed, two of the seamen, named Dudley and Stephens, killed the cabin boy, who was dying. The three seamen survived by drinking his blood and eating his flesh. They were rescued four days later and taken back to England, where Dudley and Stephens were charged with murder. The jury found that the boy would probably have died before the rescue, that the three seamen would probably have died too, and that the only way in which three of the four could have survived was by killing and eating the fourth. However, the jury did not agree that necessity dictated killing the cabin boy in preference to any of the others. A special panel of judges considered the "necessity" question, and they, too, held that there was not sufficient reason for killing the boy and that Dudley and Stephens were guilty of murder. Nevertheless, their sentences were commuted to six months' imprisonment.

Imaginary cases in which lawyers have agreed that the defence of necessity could succeed have contemplated the destruction of buildings in order to prevent the spread of fire, stealing food from a house in a remote area when a man is lost and starving, and disobeying traffic laws in order to give assistance to someone lying injured in the roadway. The defence of necessity in a variety of forms is known in England, Germany, India, Cyprus, Australia, and the United States. The model penal code of the American Law Institute, the draft federal code now before the U.S. Congress, and the penal law of New York State all provide for it in different ways, each basing its rules on the proposition that conduct which otherwise would be criminal may be justified by the necessity of avoiding greater harm.

The defence of necessity could not have been raised in *McFall* v *Shimp*, even assuming it to have been legally available in Pennsylvania, because the proceedings were brought in an attempt to obtain judicial permission to carry out compulsory tissue removal in the future. "Necessity" is only available as a defence in legal proceedings brought after the performance of a criminal act. In the case of the dying President, if this defence were available in the state where the accident occurred, it might well tip the scales in favour of the use of force upon the unwilling donor, particularly if the presidential advisers were also familiar with the minor-incompetent decisions on tissue removal, McFall's case, and the rescue principle. While the result would be a just one if this occurred, it would not augur well for personal autonomy.

CHAPTER 6

Death Redefined

I.

Which is the better source of human spare parts, the living body or the dead one? On purely medical and logistical grounds the healthy, living body in good working order is to be preferred. However, morality, ethics, and deep human considerations will always inhibit practices that reflect nothing more than these utilitarian standards.

If the initial law on the removal and use of human body materials in the 1950s had looked entirely, or even largely, to the living citizen as the preferred source of organs for pioneering transplant techniques, it is likely that public tolerance would have quickly ceased. The fact that those laws showed interest only in cadavers was the result of a deliberate decision by the lawmakers who produced them, because they must have known of the advantages of the living source. However, it soon became clear that resort to cadavers was by no means as straightforward a procedure as it might at first have seemed to be. For these early laws coincided with an historic change in our understanding of death itself. "In this world nothing can be said to be certain except death and taxes," wrote Benjamin Franklin with eighteenth-century certainty. He would not be nearly so positive if he were writing a similar letter today, because he would know that for more than twenty-five years we have been keenly aware that we cannot determine the precise moment when human life ceases. The point of divergence between the two most opposed and basic truths of existence, life and death, is not exactly identifiable, as we assumed in the past. Death is not what it used to be.

The invention of two medical machines, the ventilator and the respirator, plus the skill of the transplant surgeon, have forced us to change our perception of death, and have illuminated our inability to say unequivocally what death is, or, conversely, what life is. Fortunately we are not here concerned with what constitutes the beginning of life, yet it must be that our perception of this, too, is about to undergo a change. Since the birth of the world's first "test-tube" baby in England in 1978, the question has been at large. We must now accept and consider the implications for mankind of the proven ability to fertilize a human egg outside the human body.

The dead body nearest in condition to the living body is the one that has suffered "brain death." It will, typically, belong to a person who has received a violent blow to the head in an automobile accident or criminal attack, or has suffered a cranial hemorrhage. The victim will have severe head injuries and brain damage, and will be unconscious or in a coma, so that breathing and circulation of the blood can be maintained only by support machinery such as an artificial respirator or a ventilator. If the patient is young and healthy, and dies from causes unrelated to the condition of the organ in question, there is no more suitable source for an organ for transplant.

With patients of this kind, the diagnosis of death can be extremely difficult. Frequently they are brought unconscious into the hospital, and connected to support machinery as a matter of precaution or routine. The same thing can happen when unconsciousness develops after hospital admission, so that it is not apparent whether the patient is capable of continued, spontaneous breathing. If a person can breathe spontaneously and maintain his own blood circulation, one cannot claim he is dead, even if he remains unconscious or has suffered brain damage from which he may never recover. On the other hand, if he has suffered destruction of the brain (which may not be visible) and is incapable of spontaneous respiration, there can be proper basis for diagnosing death, even though the blood circulation and breathing are kept going by machines.

The horror of these cases and the heavy burden of decision they place upon medical staff are often worsened by the fact that the patient may have a normal appearance. The following description could apply to many cases. A young woman is a passenger in a car that collides with another head-on. She is not wearing a seat belt, so her head strikes the dashboard or windshield with great force, causing brain damage. Apart from some bruising, she has no other visible injuries. She is unconscious upon admission to the hospital, and is put in the intensive care unit with an array of tubes and similar connections to her body to provide nourishment and maintain bodily function. She is also connected to life-support machinery to ensure regular breathing, heartbeat, and blood circulation. In fact, her brain function entirely ceases within thirty-six hours, for her brain was irreparably damaged in the crash and deteriorated from that moment. Despite total loss of brain function, the support machinery ensures that the young woman continues to look as she would if she were asleep. Her body, except for bruising, has its normal appearance, the chest rises and falls rhythmically, the skin is warm and a normal colour, and she gives the impression that she could awake at any moment. In fact she is dead. Her brain is destroyed, and if the machines were switched off, her body would be unable to continue to breathe, her heart would cease to pump, and her blood would stop circulating.

Any witness could be forgiven for refusing to believe that the young woman is dead. It is not surprising that doctors themselves often find it difficult or impossible to accept immediately the doleful message that their training gives them, and "turn the switch." Disconnection of support machinery in these circumstances is a disturbing procedure which requires great mental and emotional preparation by all concerned.

There should, however, be no doubt on anybody's part that human death can, and does, occur in persons whose hearts are beating and whose blood is circulating. While not yet officially endorsed by every community, this is now widely accepted as a fact throughout the world, not only by governments, but also by churches, philosophers, and the general population.

In the past, there was usually no need for doctors or anyone else to think about death as a concept. Everybody knew that there was a near-infinite gap between life and death which, nevertheless, could be crossed in a moment. Life and death were opposites.

Any movie buff knows the old scenes where a mirror was held to the mouth of a dying man. Lack of fogging proved death. But this criterion has been outdated since the time when medicine learned that it is feasible to revive certain people who are not breathing. Following that discovery, the human spirit moved from the breath (*spiritus*) to the beating heart, which, together with blood circulation, became the focus of differentiating between life and death. This is thought by some to have been relatively recent, because the functions of the heart were not understood until 1628 when the great English physician and physiologist William Harvey published in Germany his classic Latin text *Exercitatio Anatomica de Motu Cordis et Sanguinis*. Described as the founder of modern medicine and the equal of Hippocrates, Harvey was the discoverer of the way in which the blood circulates through the body and was the first to see the heart as a pump. Most people living today have been brought up in this tradition.

It has long been known that if part of a man's heart is deprived of circulating, oxygenated blood, he may lose that part but still survive. Similarly, if some parts of the brain are deprived of such circulating blood, he may lose them too and survive. However, with the brain, the lost functions have a unique value because they are more closely identified with humanness and the "self," and include such capacities as thought, memory, knowledge, love, and learning ability. This has led to an identification of the brain with the "person" and the opinion that when the brain is destroyed, the "person" no longer exists. In other words, what we regard as making up a "person" is associated with the head, or more specifically, the brain.

Before the invention of life-support machinery any person unable spontaneously to maintain breathing and heartbeat would inevitably and rapidly die. Now, because of the very efficiency of this machinery, patients who are in reality dead exhibit a number of the characteristics of the living. Possibly the most dramatic illustration of this is the person who has been guillotined. It is scientifically possible, after the guillotine has beheaded a man, to make connections between the exposed blood vessels and then to keep the heart, lungs, and blood circulation in the trunk going for days. The headless trunk of such a victim would have the same "normal" appearance as the trunk of the car-accident victim described earlier, but no one could claim that it was alive, or not dead. Other forms of capital punishment rely on destruction of brain function as the primary means of causing death—hanging, garroting, and cyanide poisoning—although traditionally the condemned person's death has not been certified until the heart ceases to beat, which can take many more minutes.

Most people know that parts of our bodies remain "alive" after we die. Hair and fingernails continue to grow for considerable periods following the permanent cessation of heartbeat, and a long time passes before every vestige of life leaves all the tissues and cells of a corpse. The meat we buy from the butcher or the supermarket contains live cells. One of the most-repeated illustrations of the continuation of human life in cellular form concerns a woman who, years ago, gave cells from her cervix for scientific tissue culture and subsequent distribution. In the words of one writer, "to push it to its ridiculous conclusion, the lady ... is now living all over the world in glass bottles in every laboratory that grows tissue cultures."

The fact is that the death of the body is a continuing process that goes on for some time after a human being is properly regarded as dead by any standard. There is a great difference between the questions "Is he dead?" and "Is life extinct in every part of his body?" The diagnosis of death really involves a decision that a person's progress from a state of living to a state of nonliving is sufficiently far advanced to be regarded as irreversible. Normally this is a task for a medical expert. Of course, expertise is not always needed—for example, in a case of violent death involving extensive destruction of the body. Death as a "continuum" would have no application to the body of the person who was sitting on a bundle of dynamite when it exploded, or the person upon whom a one-ton slab of concrete fell. In more usual cases, however, the diagnostic example did apply until the 1960s. The traditional, classic criteria of death were permanent cessation of heartbeat, blood circulation, and respiration. These are still used in most cases. The doctor who diagnoses death by these criteria determines that a decisive stage has been reached in the *process* toward the death of every part and tissue of the body. Diagnosis by reference to cessation of brain function is no different in principle and involves a similar, but often earlier, determination. What is important is that the tests or criteria applied are scientifically accurate and that the process of dying has become irreversible.

Whether we like it or not, medical advances have brought us to the stage where hospitals frequently contain patients who, for all practical purposes, are in the same condition as the guillotine victim just described. A person can suffer destruction of the brain, even its liquefaction, and yet the remainder of his body can continue to function with support machinery. In such cases, not only is it wrong to refuse to face a distressing truth, but great injustice can be done to the patient's family, the hospital, the nursing staff, other sick and waiting patients, and the community footing the bill by insisting upon, or even allowing, the continuation of a macabre ritual of intensive medical attention upon a corpse.

The condition is nowadays commonly called "brain death." It is a description of death of the person, not of the brain as distinct from the rest of the person. A better way to describe this form of dying would be "death by reference to cessation of brain function," so as to make a contrast with other diagnostic methods such as "death by reference to cessation of respiration and blood circulation." "Brain death" refers to cessation of all brain function, including brain-stem function. A person who suffers brain death, irrespective of the condition of the remainder of his body, can never again have consciousness, memory, knowledge, thought, feeling, sight, hearing, touch, speech, or any other sense.

The profound problems associated with coma patients and deeply unconscious persons connected to artificial respirators began to attract public attention in the 1950s. In November 1957, Pope Pius XII provided guidelines for Roman Catholics in his address to the delegates to the International Congress of Anaesthesiologists in Rome. A series of three questions had been put to the Church of Rome by Dr. Bruno Haid, chief of the anaesthesia section of the surgery clinic of the University of Innsbruck, concerning the continuation of medical care upon brain-damaged patients with blood circulation and breathing artificially maintained.

The first question was whether a doctor should be obliged to use life-support machinery with every patient who could possibly benefit from it, including those whose condition, in his judgement, was hopeless from the beginning. The answer was that while all people have both the right and the duty to try to help the seriously ill, the duty extends only to the use of "ordinary means" of treatment and not to the use of extraordinary measures. "Ordinary" means of medical treatment will be determined by time, place, and circumstances of the case, but neither doctors nor anybody else should be subjected to duties of care toward others that are too burdensome. On the other hand, anybody has the right to take more steps than are strictly necessary to preserve life and health, as long as by doing so he does not neglect some more serious duty. Therefore a doctor has the right but not the duty to use life-support machinery on any patient if he honestly decides that he should.

Along with some directly religious additional concerns about the last sacraments, the second question asked whether a doctor (or other responsible person) has the right or duty to switch off life-support machinery if an unconscious patient fails to improve after some time. Where does the doctor stand in relation to a patient with heartbeat and blood circulation who will be unable to breath spontaneously and will quickly perish after the switch is turned off? The Pope's answer was that a doctor may exercise his own judgement and disconnect the artificial support. Because it is "extraordinary treatment" to start with and is therefore not morally obligatory, artificial support may be discontinued without breach of duty or moral obligation.

As for a patient's family, the Church is of the view that they are not bound to agree to anything more than "ordinary means" of treatment. The family would therefore lack the right to insist that a doctor use life-support machinery, and presumably also the power to prevent him from using it if he decided to do so as a matter of medical judgement. If, however, it could be shown that the patient himself was opposed to "extraordinary" measures, neither the doctors nor the family may overrule his wishes. As to continuation of extraordinary treatment that fails to produce improvement, the family have the right of veto after a reasonable period, and the doctor should comply—though they would not have the right to insist that the treatment continue against a contrary medical judgement. The Church's opinion here rests upon the proposition that turning off life-support machinery does not involve a "direct disposal of the life of the patient, nor . . . euthanasia in any way: this would never be licit."

The third, and on the face of it the most difficult, question was whether coma patients whose "lives" are sustained by support machinery should be considered to be dead once it has been determined medically that they are irrecoverable. Put the opposite way: Is the patient who breathes and has blood circulation, even though these depend entirely on artificial apparatus, to be regarded as alive until those bodily functions cease, whether the cessation is caused by turning a switch or by total systemic disintegration? To some, the Pope's answer was surprising. He said that the decision as to whether death has occurred in a human being, and the selection of the time of its occurrence, is a medical matter. Responsibility for declaring the fact of death rest upon the doctors. The Church comes into the picture both before and after death, with rituals, services, and beliefs that revolve around the fact, but it is not the Church's role to diagnose or define death. Definition of death is also for medical experts. The Pope's words were, "verification of the fact [of death] . . . cannot be deduced from any religious or moral principle and, under this aspect, does not fall within the competence of the Church." He agreed that the coma patient presents enormous problems: "A great number of these cases are the object of insoluble doubt."

The Roman Catholic Church thus provided a lead at an early stage by confirming that the diagnosis of death and the prescribing of criteria to be used, whatever bodily function the criteria may refer to, are medical responsibilities. The Pope's answers provided a coherent basis for regulating an important new field of medicine, but as sensible as his precepts are, they do not conform to the secular laws of every society. The doctor who follows them should check his legal position.

After the Pope's pioneering pronouncement, the subject of brain death retreated from the forefront of international consciousness for some years until the furor caused by the first successful heart transplant in December 1967. It was as though, with the realization that organ transplantation was here to stay and would extend into unknown areas, the world of medicine, and governments everywhere, received a sudden injection of intellectual and political adrenaline. It needed little intelligence to appreciate that transplant surgeons and their patients would use every available opportunity to obtain beating hearts and functioning organs from brain-dead cadavers.

Not only was there no global medical consensus or criteria for diagnosing brain death, but there were daunting difficulties of public acceptance, and serious legal problems for transplanters under the law as it then stood. As for legal problems, the simplest way of putting it was to say that, unless there was a change in laws that only recognized death by reference to cessation of heartbeat and blood circulation, a surgeon who removed a functioning vital organ would probably be guilty of homicide.

Public unease would be one of the least surprising responses to signs that a change is about to be made in the official perception of death, and that people will be declared dead tomorrow who cannot be declared dead today.

People's reactions to sensitive questions about death and dying cannot always be anticipated. From time to time, they have manifested themselves unexpectedly and suddenly. We have already traced the story of the British anatomy schools and the suppression of body-snatching in that country, but it should not be forgotten that American opinion was no less inflammable. In April 1788, five people were killed in a riot in downtown New York following a mob burning of the Hospital Society's building because it housed an anatomy collection. The rioters would have "no fooling with human bodies." In the nineteenth century, both Britain and the United States saw the growth of a bizarre public preoccupation with premature burial. Examples were widely publicized of apparently dead people regaining consciousness as their coffins were about to be lowered into the grave, particularly after drowning or electrocution. This was seized upon as a theme by writers, one of the most prominent being Edgar Allan Poe. The public itself came up with a variety of solutions, one of the most obvious and unpleasant being simply to delay burial until corpses began to decompose or at least to smell strongly. Another was to produce an amazing selection of mechanical contrivances for attachment to corpses and coffins—warning and signaling devices whereby a buried person could make his recovery known. Cords, wires, levers, springs, pipes, ladders, mirrors, air circulators, alarms, and many other things were tied to bodies, led through coffins to the open air, or placed in vaults to enable rescue and periodic inspection. Attendants were hired to sit in graveyards, and food and tools were often placed inside coffins. Spring-loaded coffins in the vaults of the rich had the habit of flying open long after burial because of mechanical defects or vibration. The obsession did not completely die out until well into the twentieth century, and

the *American Journal of the History of Medicine and Allied Sciences*
lists twenty-two separate inventions of this kind fully patented in the
United States between 1868 and 1925. All of this proves that the diagnosis of death in certain unusual cases has long been a matter of great
difficulty.

A quarter-century ago, public unease manifested itself simultaneously with the early publicity attending the appearance of patients
in permanent coma who could breathe only with artificial aid. Such
patients were novel, and could never have "lived" until the advent of
support machinery. Such a patient, whose brain function has ceased
and whose body is entirely dependent on external equipment for the
maintenance of its blood circulation and heartbeat, is quite different
from a patient who is in a coma but whose body functions spontaneously. The difference is fundamental. Patients have lain unconscious
for years without being dead. Possibly the longest known case of this
kind was a woman who died in the United States in 1978 at age forty-
three, after having been in a coma for more than thirty-seven years following an operation to remove her appendix.

With life-support machinery, the necessity to declare death in a
brain-dead person was soon apparent to the medical profession, but it
was not so apparent to the public at large. The doctors needed to replace centuries of practice and a lot of ignorance with new standards,
so that they could classify as dead patients who were not dead under
existing standards. The fact that it had never before been possible to
have patients of this kind—that they were "made" by modern medicine—did little to allay the disquiet of those who saw the proposal as a
man-made rearrangement of life and death.

As it has turned out, the most disquiet was manifested among small
groups of politicians, lawyers, and medical practitioners themselves.
This is not surprising, as these groups have had more to do with the
concept of brain death than others, and have had to decide upon action or inaction with little precedent or guidance.

One of the most persistent explorations of public feeling on brain
death occurred in 1976–1977 in Australia, when the federal Law Reform Commission produced a model statute on transplantation, anatomy, and autopsy. Because of the relationship between brain death
and organ transplantation, it was decided that a direct assessment of

public opinion was necessary. A Gallup Poll conducted in November 1975 had disclosed that 71 percent of the population believed that irre-coverable coma patients should not be kept on artificial machinery. The official inquiry included public hearings in every capital city at which any member of the public was entitled to voice an opinion. Prior to the public hearings, a working paper on the issues was circulated to governments, churches, medical associations, universities, community groups, television, radio, the press, and individuals known to have ex-pertise. The press, radio, and television gave extensive coverage.

In its report, the Law Reform Commission announced that the public discussion had produced "a clear indication of public accep-tance." Not one witness who appeared in the public hearings disagreed with the concept or with the idea of its enactment into law. With very few exceptions, the written submission from the public expressed the same attitude. "The procedures followed, and the wide media public-ity given to 'brain death' would have produced evidence of public anx-iety and apprehension if such existed. There has been none. The reservations expressed . . . have been few. . . ." It was also found that public unease, when it did appear, was from lack of information. Pro-vided with a clear and intelligent exposition of the subject, participants in public meetings and members of the public at large showed equal clarity and intelligence in their responses.

Following this inquiry, a model uniform statute was promulgated that recognized the concept of brain death as an addition to the classic means of diagnosing death by reference to the cessation of blood cir-culation.

Without a legal recognition of brain death, extremely unpleasant consequences can overtake not only transplant surgeons but the inten-sive care expert who "pulls the plug" to discontinue life-support ma-chinery. These consequences range from damage suits to homicide charges, and although many countries have eradicated the problem by means of specific statutory provision, many have not. In many places, confusion has ruled because the law has defined death only in the most general terms or has not defined it at all.

Numbers of examples from a variety of countries can be given of doctors who have fallen foul of the law by accepting and acting upon the correctness of the brain-death concept. The head of the surgery department of a prominent Japanese university medical school remained for more than two years under threat of criminal charges of double homicide after performing Japan's first heart transplant in August 1968. The donor was a twenty-one-year old student who had drowned, and the recipient a youth of eighteen who had a malfunction of a main heart valve and other complications. Both the local prosecutor's office and later the supreme prosecutor's office in Tokyo carried out long investigations following complaints that the surgeon had removed the donor's heart before he was dead and was guilty of "vivisection." Another allegation was that he should have tried to treat the recipient by other means than transplantation. The lack of a legal definition of death in Japan complicated the considerations. In September 1970, the supreme prosecutor's office ruled that there was insufficient evidence to bring homicide charges. In 1965, a Swedish surgeon who had removed a kidney in comparable circumstances was accused of "contributing" to the donor's death. This triggered a nationwide debate, which, incidentally, indicated a favourable view of his action.

In the United States, there has been a variety of legal proceedings against transplant surgeons. One of the most widely publicized is known as the Richmond Brain Case (proper title, *Tucker's Administrator* v *Lower*), a civil trial of fourteen surgeons and physicians in the Law and Equity Court of Richmond, Virginia. This was an action brought by the brother of a brain-dead patient at the Virginia Medical College Hospital, claiming $1 million damages for wrongful death. The trial commenced on May 18, 1972, and the jury of seven gave its verdict a week later.

The deceased was a fifty-six-year-old man, Bruce Tucker, who tripped at his local gas station and struck his head on a concrete apron late in the afternoon of May 24, 1968. He was able to get to his feet and walk away, but an hour or so later had to be taken to the hospital's emergency room. By 11:00 P.M., his comatose condition was such that the surgeons opened his skull to try to relieve brain hemorrhage and pressure from swelling. A little over twelve hours later, at 11:30 A.M. on May 25, he was placed on an artificial respirator, and within a few minutes the attending physician wrote on his chart "prognosis for recovery is nil and death imminent." Tests were carried out over the next hour or so, as a result of which a staff neurologist concluded that the patient's brain had irreversibly ceased to function and that he was dead. He was kept on the respirator and later that afternoon his heart was removed and transplanted to another patient.

During the morning, one of the medical team attending Tucker advised hospital staff to try to locate his family to obtain permission to use his heart and kidneys for transplantation. Both before and after this request other members of the team notified the police in an attempt to contact the family. The first message to the police had been given at midnight, and the second at 2:00 P.M. on May 25. The police phoned back thirty minutes after the second call saying that they had been unable to reach any next of kin. By a sad irony, a friend of the patient had been unsuccessfully "roaming the corridors" of the hospital buildings from midday to 3:00 P.M. trying to find him. In addition, a business card of Tucker's brother that was in Tucker's wallet was either not seen or not acted upon by the hospital. It was this lack of notification and absence of family consent that probably caused the brother to file suit.

The brother claimed that the law of Virginia at that time did not recognize brain death, and had regard only to the traditional criteria of death. It was therefore contended that the doctors were wrong to conclude that Tucker was dead when his brain function was shown to have stopped permanently. According to the plaintiff's argument, Tucker was still legally alive at the time his heart was removed from his body, because he still had a heartbeat, respiration (admittedly maintained by machinery), blood pressure, and body temperature.

The jury needed only seventy-seven minutes to accept the concept of "brain death" and to return a verdict in favour of the doctors. One juryman, in a later interview, was reported to have said, "It was clearly proved in the trial [that] a man . . . cannot live without a functioning brain." Another is reported to have declared that the traditional legal definition of death was no longer acceptable. The Virginia legislature agreed with the jury, and in the 1973 session of the Assembly enacted a statute that put the matter beyond doubt, and legalized the use of brain-death criteria in diagnosing death.

Cases similar to the Richmond Brain Case continue to appear in states where the law has not been clarified. In June 1978, a Minnesota judge warily authorized the bringing of a murder charge against a mother who had allegedly assaulted her four-year-old daughter. The child had been declared dead on brain-death grounds, but was being maintained on a respirator because the hospital doctors feared to turn it off. The judge said that the case raised "a very sensitive" question because the definition of death had not been reduced to clear terms. The issue of a death certificate by a doctor on brain-death grounds did not fulfil all the requirements of Minnesota law, but he felt he should accept it "to some extent."

In states and countries that have no statutory recognition of brain death, transplant surgeons often fear that they will be involved in homicide trials, in the manner that befell Dr. Norman Shumway in a California case in May 1974. Dr. Shumway, one of the world's best-known heart-transplant surgeons, had removed a heart from a man who had been shot in the head. The assailant's defence in his criminal trial was that the victim's death was caused not by the bullet but by Dr. Shumway. At the trial Shumway flatly rejected this and said, "The brain in the 1970s and in the light of modern day technology is the *sine qua non*—the criterion for death. I'm saying anyone whose brain is dead is dead. It is the one determinant that would be universally applicable, because the brain is the one organ that can't be transplanted."* The jury accepted this statement and found the accused guilty of voluntary manslaughter. California gave statutory recognition to brain death four months later, in September 1974.

Confusion has similarly prevailed in England, as illustrated by the rare cases that have reached the English courts. In England, the law has defined death only in the most general terms, or not at all. Some lawyers take the view that there is no legal definition of death, while others offer loose definitions such as "the absence of vital functions." However, no laws have been enacted to resolve the doubts. The first case was a coroner's inquest into the death of John David Potter in Newcastle upon Tyne in July 1963. Potter, a thirty-two-year-old fitter,

* In the light of present and foreseeable technology, it is obviously correct to say that brain transplants are an impossibility. Yet experimental work never ceases. *Science* magazine in December 1980 carried a report that United States scientists are rapidly developing the capacity to correct deficiencies in the central nervous systems of animals by transplanting brain tissue. Such transplants between rats have successfully cured a type of diabetes and have ameliorated some forms of brain damage.

had been in a fight the previous month and had fallen backward on his head after being butted twice in the face. He was taken unconscious to the hospital, where he was found to have four fractures of the skull and extensive brain damage. Fourteen hours later, his breathing stopped. The hospital doctors then decided to connect him to a respirator, although they had formed the opinion that he was dead when his heart had ceased beating at the time his breathing stopped. The respirator restarted the breathing process and the doctors then asked Potter's wife for consent to remove one of his kidneys for transplant. She agreed, and while he was still on the respirator, a kidney was taken and successfully transplanted to a waiting recipient. After the kidney was removed, the respirator was turned off, and there was no spontaneous breathing or circulation. At the coroner's inquest, the question arose whether Potter's assailant had caused his death, or whether it had been due to the removal of the kidney or turning off the respirator.

One medical witness expressed the opinion that Potter did not die until the heart ceased beating and the blood ceased circulating after the respirator was disconnected. Two other doctors, one of whom was a brain surgeon, gave contrary opinions, to the effect that the brain damage had caused the death and that the kidney removal had nothing to do with it. The coroner's jury followed the brain-death opinions and decided that the kidney removal had not caused Potter to die, nor had the disconnection of the respirator. A verdict of manslaughter was returned against Potter's assailant. However, in the subsequent criminal proceedings, the accused was found guilty only of common assault, and not of manslaughter. Some legal analysts believe that this result came about because the prosecuting lawyers were uncertain of the legal validity of the jury's verdict at the inquest.

Published reports of Potter's inquest are insufficiently detailed to allay concerns of lawyers and coroners in England and elsewhere. The worry has persisted that in comparable cases the surgeon is at legal risk. If an assailant is charged with murder or other homicide, it may be open to him to argue that the victim's death was caused not by him but by the surgeon who removed an organ or turned off a respirator. The defence could claim that the concept of "brain death" is not legally accepted and that the traditional criteria of death are the only correct ones.

Such was the effect of the Potter proceedings that in March 1977 a heart transplant was cancelled in Melbourne, Australia, where the law was similar to the English law, solely because of a coroner's legal doubts. The prospective donor was a young man who had suffered brain death as a result of bullet wounds, but because of the possibility of criminal charges, the coroner expressed the view, based on Potter's case, that it was legally unsafe for the surgeons to proceed because they might give the assailant a ready-made defence. Despite the fact that the donor's family wished his heart to be given and that there was a waiting recipient, the operation was cancelled. The proposed recipient died of heart disease shortly afterward.

It was not until fifteen years after *Potter* that England saw the appearance of cases indicating a more positive legal attitude toward brain death. This was long after the medical profession and the public in England and in the West generally had accepted the concept. By 1978, many Western communities had enacted laws that recognized the concept, but England had not, despite the unanimous formulation by all its Royal Colleges of Medicine in 1976 of brain-death criteria which were accepted around the world as authoritative. In 1978 and 1979, three cases in England lent support to the use of brain-death criteria as a lawful means of diagnosing death, but unfortunately, none of the decisions was made by a superior court of record, and until that occurs, England will continue to have no binding or final statement of law and doctors will still be involved in legal proceedings.

The first case occurred in February 1978 when a seventeen-year-old youth was kicked unconscious outside a discotheque in a gang attack in Norwich. After conducting a series of "brain-death" tests, the doctors in the hospital to which he had been taken decided that he was dead and switched off the life-support apparatus with his father's consent. There was no subsequent legal criticism of the doctors, although one member of Parliament took the opportunity to describe their action as a "blatant case of euthanasia." A few days later, a coroner's jury reached a verdict similiar to that which the Potter jury had given.

In the second case, an inquest investigated the death in October 1977 of one Carol Wilkinson, a woman of twenty, who had been violently attacked on her way to work in Bradford. Admitted to the hospital on Monday, October 10, she was immediately put on a ventilator when it was found that she had severe head injuries. She was continuously monitored and tested for three days. At no time was there any detectable brain function. After a final series of tests on Wednesday, October 12, the doctors decided that she had suffered brain death and that the ventilator should be withdrawn. With the agreement of Carol's parents, the coroner, and the police, this was promptly done. At the coroner's inquest, the jury attributed responsibility for her death to her attacker and decided that she had been "unlawfully killed by a person or persons unknown."

Eighteen months after Carol Wilkinson's death, a twenty-two-year-old gardener named Anthony Steel was charged with her murder. Steel denied the murder charge, although he admitted the attack. At the trial, which was not completed until December 1979 at Leeds Crown Court, the defence contended that the fact that the doctor in charge of the case had turned off the life-support machinery raised the question whether Carol's death was really due to the attack. The judge ruled that the doctor's action did not contribute to the death and was not relevant to the trial. The jury returned a verdict of guilty and, on December 13, Steel was sentenced to life imprisonment for murder. It was promptly announced that his lawyers were considering an appeal, although it was not stated whether the brain-death question would be raised.

In November 1979, Richard Malcherek, a thirty-two-year-old electrician, was tried at Winchester Crown Court in Hampshire for murdering his wife by stabbing her eight times with a kitchen knife. The attack took place about midnight on March 26, 1979. Christine Malcherek was taken to the hospital, where she appeared to be doing well, but on April 1 she collapsed and was transferred to an intensive care ward. Shortly after midnight, she lost consciousness and her heart stopped. After emergency surgery to remove a blood clot from her lung, and after heart massage, her heartbeat resumed, but her brain had been deprived of oxygen for twenty-five minutes. She was connected to a ventilator but did not improve, and by April 5 the doctors decided that she had suffered brain death. The ventilator was turned off. At the trial, the prosecuting counsel told the jury it would have to decide whether Mrs. Malcherek was murdered by her husband or died as a result of the doctors' having disconnected the ventilator. "Who caused that woman's death? Surely you would not say that doctors did it. In the submission of the Crown that would be an absurdity," the prosecutor said. Malcherek denied murdering his wife. The judge, like the one in Carol Wilkinson's case, found that "there had been no evidence to show that Mrs. Malcherek was not dead before doctors switched off the machine." The jury took under an hour to return a verdict of guilty. Malcherek was convicted of murder and sentenced to life imprisonment on November 13, 1979.

Some specialist lawyers in England take the view that the diagnosis of death by reference to cessation of brain function is accepted by common law. But their clear vision is not shared by everybody, including the prosecution and the defence in the cases of Malcherek and Steel. Until it is, or until Parliament or an appellate court clarifies the law, doubt will remain about the legality of brain death and the transplant of organs. It is damaging to the public interest for legislatures to allow these doubts to continue and cause needless deaths when they could be quickly dispelled by a brief statute. The Melbourne case in 1977 was a demonstration of needless death. Nobody will ever know how many other lives have been lost because transplant operations have not taken place due to legal fears, or how much subterfuge such fears have forced upon medical practitioners in the withdrawal of artificial life support, or even in the reluctance to use it in the first place.

Ironically, there are no more positive medical guidelines for diagnosing brain death than those that exist in England, and there is general agreement that the English public accepts the concept. Statements to this effect are made with increasing frequency, and emanate from such conservative office holders and instrumentalities as the chairman of a government working party on brain death and transplantation, the president of the Royal Society of Medicine, prominent coroners, a government department, and *The Times* of London. At the same time, successive English governments, with full knowledge of these facts, have indicated that they have no intention of recognizing brain death legislatively.

A strange, incomplete report from Florida in November 1975 gives further reason for conjecture on the impact of legal doubts upon medical practitioners. A sixteen-year-old girl with a terminal illness was being maintained on an artificial respirator in Nassau Hospital when the plug of the respirator was pulled out of the wall by a "person un-known." Complete respiratory failure and cardiac arrest promptly fol-lowed the disconnection. The coroner was unable to ascertain who pulled the plug, whether hospital staff, family visitors to the patient, or anyone else. However, he took little time to rule that the patient had been the victim of homicide, probably murder. The ruling shows that in intensive care medicine, the legal risks can be very high. A doctor who is uncertain of his legal rights and duties in a field where people talk of criminal homicide charges may well refuse or omit to begin treatment, or he may feel compelled to be secretive or overly cautious about its continuation or cessation.

The law should speak, and speak plainly. It is not good enough to write rules on a kind of juristic Rosetta Stone that can be read only by skilled interpreters. Whether they are promulgated by the courts or the legislature does not matter—although the legislature would be a faster and better vehicle for progress. A sidelight showing how history can be changed by legal confusion was provided by Dr. Christiaan Barnard in a public discussion in November 1979. He was asked why it was that he came to perform in South Africa the world's first successful heart transplant in 1967. His answer was that American transplant surgeons had been in a position to do the operation for some time, but had been deliberately holding back because they were afraid of their legal liability.

II.

Traditional law has been tentative in defining death and slow to recog-nize the concept of brain death. This has caused many problems, but in the overall picture the problems have been relatively minor. There have been far more important obstacles to surmount—for example, the need for the medical profession itself to be satisfied that permanent cessation of brain function *is* a true gauge of death, and the develop-ment of medical criteria enabling a positive diagnosis.

The story of the medical profession's struggle to achieve medically acceptable criteria for brain death and to give affirmative answers ca-pable of translation into modern laws can be conveniently commenced in 1968.

In August 1968, the Twenty-second World Medical Assembly in Sydney, Australia, proclaimed the Declaration of Sydney. This Declaration, of some three hundred words, dealt solely with the determination of death, and in particular the time of death. After noting that the classical criteria of death would still be valid for most cases, the Declaration drew attention to the changes in diagnosing death brought about by artificial life-support machinery and advances in organ transplantation. It also emphasized that death is a gradual process, with different tissues and cells dying at different times.

> But clinical interest lies not in the state of preservation of isolated cells but in the fate of a person. Here the point of death of the different cells and organs is not so important as the certainty that the process has become irreversible. . . . Determination of the point of death of the person makes it ethically permissible to cease attempts at resuscitation and in countries where the law permits, to remove organs from the cadaver.

Earlier that same month, the *Journal of the American Medical Association* had published a historic report by a group of Harvard Medical School specialists. The Harvard Report opened with these words: "Our primary purpose is to define irreversible coma as a new criterion for death." It then listed a series of tests whose application would enable an accurate determination of whether brain death had occurred. "The Harvard criteria" called for detailed verification of lack of responsiveness, movement, reflexes, spontaneous breathing, and brain function. They included use of the electroencephalograph (EEG). If all the tests showed negative responses, this would confirm irreversible loss of brain function and the patient's death. Unhesitatingly the report said that death should be declared *before* discontinuing artificial life support. "It should be emphasized that we recommend the patient be declared dead before any effort is made to take him off a respirator. . . ."

The Harvard Report was broadly, although not universally, accepted. Many experts were concerned about the reliability of the EEG, and a continuing debate took place on that subject. (Eventually the use of the EEG dropped from favour, as will be seen from the 1976 statement of the British Royal Colleges.) Even so, the Harvard Report, especially its comprehensive clinical criteria (as distinct from the EEG), was endorsed by an eminent A.M.A. task force four years later in July 1972. The Harvard criteria, or something very much like them, found their way into statutory definitions of death in Europe, for example, in Italy and Switzerland.

A few months before the Report, on April 24, 1968, the French Ministry of Health issued a circular establishing new rules and criteria for the declaration of death. The decree prescribed a series of tests similar in range to the Harvard criteria, and forbade a determination of death until they had all supplied the appropriate negative answers.

In November 1968, the Canadian Medical Association published brain-death criteria that were clearly inspired by the Declaration of Sydney and the Harvard Report. These were succeeded in 1974 by a more detailed Canadian definition.

Both the acceptability of brain death as a standard of death and the criteria to be used were given powerful support and updated by a unanimous public statement of the Medical Royal Colleges and their faculties in England on October 11, 1976: "It is agreed that permanent functional death of the brain stem constitutes brain death and that once this has occurred further artificial support is fruitless and should be withdrawn." The statement drew attention to the preceding period of philosophical argument on the diagnosis of death, "which has throughout history been accepted as having occurred when the vital functions of respiration and circulation have ceased," and paid a compliment to the Harvard Report. The new guidelines, the fruit of collected medical experience in the period between 1968 and 1976, were declared to be sufficient to distinguish between patients with even a remote chance of recovery and "those in whom no such possibility exists." The prescribed tests were designed to confirm coma, irremedial structural brain damage, absence of all brain-stem reflexes, and inability to breathe spontaneously. The tests were to be repeated at intervals, together with other procedures. (The statement specifically excluded the use of the EEG.) In January 1979, the Royal Colleges published a further memorandum to eliminate any doubt that brain death amounted to death for all purposes. The closing words of this document were "brain death means that the patient is dead, whether or not the function of some organs, such as a heart beat, is still maintained by artificial means."

In the light of such continuous authoritative medical pioneering, it is not surprising that the more active societies soon began to create or enact laws directly recognizing brain death. The first European step was taken by the French, in April 1968, not with a public law but with a departmental circular issued pursuant to the decree of 1947, which had given a number of listed hospitals powers over dead bodies. Still, it was a beginning, and in 1969 was followed by a Swiss statement. The Swiss Academy of Medical Sciences published in that year their directions for diagnosing brain death. In 1971, the Canton of Zurich adopted these as an Ordinance, and they remained in force after the defeat in 1972 of an appeal to the Swiss Federal Court by opponents of the concept.

Slowly the lawmaking gathered impetus throughout the Western world, so that in 1978 the Council of Europe gave direct recognition to brain death in its model rule that designates the time when body parts may be removed from a cadaver: "Death having occurred, a removal may be effected *even if the function of some organ other than the brain may be artificially preserved.*" (Italics added.)

In the United States, the spread of state statutes on brain death has been continuous since Kansas enacted the first in 1971. An imprimatur of general acceptance was given in July 1978 by the Commissioners on Uniform State Laws. We have already seen how the Uniform Anatomical Gift Act of 1968 came into being; a decade later came the Uniform Brain Death Act. In a "prefatory note," the commissioners confirmed that medical practitioners are concerned about their vulnerability under traditional law when life-support systems are withdrawn. "This Act expresses community approval of withdrawing extraordinary life support systems when the whole brain has irreversibly ceased to work." The act, as brief as the Californian, is silent on the criteria to be used, leaving this to the medical profession. The definition is: "For legal and medical purposes, an individual with irreversible cessation of all functioning of the brain, including the brain stem, is dead." The act allows the traditional criteria of cessation of blood circulation, breathing, and heartbeat to be used if required. The word "functioning" is critical, and is intended to describe *purposeful activity* in all parts of the organ, as distinguished from random activity of no meaning, which can sometimes be detected by sensitive monitoring equipment in a nonfunctioning brain.

Canada dealt with the subject on a federal level in 1979 with a Working Paper by the Canadian Law Reform Commission entitled "Protection of Life: Criteria for the Determination of Death." This document recommended accepting that death may be lawfully diagnosed by reference to cessation of brain function.

Acceptance by means of legislation is now sufficiently broad to suggest that laws on the subject will become general in due course, as the more conservative or timorous governments take action. The latest laws, such as the American Uniform Act and those in Canada and Australia, tend to be extremely brief: they simply recognize the notion of brain death and eliminate liability for doctors who use it. Unlike the early European regulations of France and Switzerland, which set out in detail the necessary criteria, the newest laws provide no more than a framework. Ordinances and circulars such as the Swiss and French can be readily changed, but with a public statute the difficulties and delays in getting parliamentary amendment make this broad phrasing preferable so that the law will not become outdated or redundant because of technological advances. The Australian model statute in 1977 was accompanied by an invitation to the medical colleges to publish medical criteria that would enable accurate diagnosis of "irreversible cessation of all function of the brain." In this way, criteria can be developed and changed from time to time as medical knowledge increases, without any need to alter the statutory definition.*

* Public concern and restiveness on the subject of death is still such that those who invite (or incite) public reaction to the concept of brain death carry an unusually heavy responsibility. One reason is the possibility of widespread public alarm. Another is that a sudden falling off of body part donations, particularly organs, must, under today's conditions, result in increased numbers of deaths among potential recipients. A well-documented example was provided by a major British television program on October 13, 1980, which suggested that current British surgical practice might have been allowing the removal of organs from patients certified as dead when those patients could recover if the organs were not removed, i.e. that the donors were not dead when organs were removed. The basis of the allegations was a series of United States (not British) case histories in which the program claimed that patients were wrongly diagnosed as having suffered brain death. No evidence was offered that such errors had actually occurred in Britain, and the British Royal Colleges of Medicine promptly denied that their 1976 brain-death criteria were in any way defective.

The immediate result was intense national and international publicity, a public estrangement between the British Medical Association and the British Broadcast-

III.

When the law recognizes and accepts brain death, what is in store for the brain-dead body? Is there a greater likelihood of dead bodies being seen as community property? The answer must be yes. The patient who is diagnosed as brain dead will be in a hospital, because it is only in a hospital that his deteriorating condition will have been arrested and the diagnostic tests carried out. Transplantation also must take place in a hospital. In absolute numbers, only a small proportion of deaths will fulfil all conditions for donor suitability. In order to serve the interests of the sick, it is necessary to enact public laws of general application to ensure that the net is spread widely enough.

The average person has little need to worry at present about claims by the sick or the community upon his body, despite the unrestricted scope of the laws relating to this subject. Yet one should not feel too secure, and even less so about the future. After all, transplantation of organs is by no means the only medical activity which requires a continuous supply of dead human bodies and body parts. The claims of experimentation, therapeutic extracts, anatomy schools, autopsies, and research should not be forgotten.

The next few pages deal with possibilities that have received serious attention from academic writers and have even been the subject of best-selling novels and movies. The first such popular novel was *Coma* by Robin Cook. To make conjecture even more credible, or for some,

---

ing Corporation, refusal of the Association to participate in further programs on the subject, condemnation by Members of Parliament in the House of Commons, and the virtual collapse of kidney transplantation in the United Kingdom in the following month. Leading transplant surgeons made their reactions known in the press. Professor R. Y. Calne wrote, "The result of this biased prosecution of British transplantation practice will be distressing and its perpetration was a wicked act." The surgeon in charge of kidney transplants at Guy's Hospital, London, was reported as saying that he was terrified at the possible effects of the program and that patients would be left as vegetables because relatives would no longer allow support machines to be switched off. *The Times* of London, in an editorial on November 26, called the program "very bad . . . indeed." In November 1980 only thirty-two kidney transplants were performed in Britain, "one-third of the normal number and the fewest for many years." The Minister for Health, in early December, said that the government's campaign to increase the numbers of transplant donors throughout the community had been "tragically affected" by the program.

alarming, it should be borne in mind that not every nation, state, or province has laws which compel the burial or cremation of the dead. Some do, but others have only general rules of public health and hygiene. We are once again in an area where, in the past, law has not really been needed. People have always mourned their own dead, and have buried them without governments forcing them to do so. What would happen if we decided *not* to bury our dead, or at least, not all of them, but instead to preserve and keep them? Some writers and thinkers on this subject have assumed that the keeping and preserving of the dead is necessarily illegal. It isn't.

The prefatory notes on the United States Uniform Brain Death Act concluded with the following words: "Some other questions and subjects not addressed by this narrow act are: . . . euthanasia . . . *maintaining 'life' support beyond brain death* . . . and protection accorded the dead body" (my italics). This means that after someone is declared dead by brain-death criteria, it is not legally obligatory to turn off life-support machinery unless a specific law mandates it. In fact, support machinery is often kept going when the patient is to be a transplant donor. What would happen if the patient was not to serve as a transplant donor, at least not immediately, but the support was kept going anyway?

In an article published in 1975 in the United States, Willard Gaylin, the President of the Institute of Society, Ethics, and the Life Sciences pursued this very question. Talking of patients who have suffered brain death, he noted that after the declaration of death, it could be lawful *not* to pull the plug, with the consequence that they would unquestionably be corpses, but utterly unlike any corpses known throughout man's previous history: they would be "warm, respiring, pulsating, evacuating, and excreting bodies requiring nursing, dietary and general grooming attention." The fittest of them could be maintained for long periods. Coining the expression "bioemporium," the author described an imaginary but feasible hospital or similar building in which these cadavers would be kept. Long, silent rows of them, looking like any other comatose patients, would be laid out for the purpose of being, in effect, farmed and harvested for the benefit of the living.

The benefits such cadavers could confer upon the community

would be enormous. As subjects for medical training they would be invaluable, saving embarrassment and discomfort for both parties as a student carried out his first physical examinations—use of the stethoscope, examinations of the eye, ear, nose, rectum, vagina, and other parts of the body. Mistakes could be made without causing injury, operations could be performed, limbs amputated, and skin grafted. Organs could be removed, whether simply for experiment or for transplantation. Testing of dangerous new drugs and new surgical procedures, experimentation with viruses, poisons, cancer cells, and so on could be done at will. Again, these cadavers could produce steady supplies of blood, bone marrow, and other tissues capable of being implanted in the sick, as well as serving as producers of antitoxins and antibodies to fight major diseases after injection with the appropriate toxin, virus, or bacteria. As a basis for cancer research and immunological research, they would be invaluable.

With total annual deaths in the United States at roughly 2 million and the number occurring from accidents, suicides, and homicides at around 275,000, Mr. Gaylin deduced that a substantial number of suitable cadavers could end up in bioemporiums, certainly enough to be of tremendous public benefit at a low cost. Since this system promises great help in finding cures for forms of cancer, including leukemia, and reducing much existing suffering, widespread public tolerance of such institutions might be anticipated in the future. The main inhibition at present would be "revulsion . . . as a bias of our education and experience, just as earlier societies were probably revolted by the startling notion of abdominal surgery, which we now take for granted."

The question is whether the revulsion we feel upon learning of the technical practicality of these activities arises from our essentially human characteristics, which we could not eliminate without diminishing our humanity, or from our current moral attitudes and training, and fear of new technology.

It will be more or less likely that the bodies of the dead will be used for strange or shocking purposes (judged by present standards), depending on the criteria used for diagnosing death and the laws on the disposal of corpses. If the criteria for determining brain death are severe, such as those spelled out in England in 1976, few brain-dead cadavers could function for more than a day or two after death is determined, even with machine support. Under these criteria, death cannot be pronounced until doctors are satisfied by repeated tests that spontaneous respiration and heartbeat are impossible. A body utterly unable to sustain heartbeat and blood circulation spontaneously can be kept going by machine only for a limited time; soon the entire nervous and physical system will begin to disintegrate, and even the machinery will not be able to maintain basic functions. In places where the criteria are not so stringent, however, it may be possible to declare death while the body has the capacity to function for a much longer period with the aid of support machinery.

As for the burial or cremation of the dead, a few communities prescribe specific duties. California's Health and Safety Code, for example, states that "the duty of interment and the liability for the reasonable costs of interment of" the remains of a deceased person devolve upon the surviving spouse, surviving children, surviving parents, and nearest kin, in that order. But laws imposing a direct duty to bury are relatively rare. Usually the subject is covered by traditional legal principles that fall short of placing direct, enforceable duties on named persons. Much energy has been devoted in medicolegal journals in England and America to analysing traditional principles; but whatever they are, obviously they may be ignored when a body is donated, as the law allows, to a medical school or research laboratory for autopsy or for removal of tissue for transplantation.

In the very first year of operation of the Anatomy Act of 1832, Jeremy Bentham, one of England's greatest political writers, reformers, and jurists, died. The inventor of the principle of "utility"—according to which the purpose of all political and social acts should be to increase pleasure and reduce pain by trying to do the greatest good for the greatest number—and the author of a torrent of books, articles, and pamphlets, he was also one of the founders of University College in London. Bentham had an amazingly fertile imagination. His reformer's eye was caught by the Anatomy Act's encouragement of body donation, so he decided to set an example. His will contained elaborate provisions for the disposition of his corpse. First it was to be given to University College's anatomy school for examination. Next, the skeleton was to be reassembled, rebuilt, and clothed in his own garments. The body, thus restored to a presentable appearance, was to be placed in a cabinet and presented to the college to be put on exhibition. To induce the fellows of the college to go along with this plan, he bequeathed a sum of money to pay for an annual dinner for them to honour his memory. To this day, the body of Jeremy Bentham sits in the cloisters just inside the ground-floor entrance of University College in Gower Street. It wears a large-brimmed hat and is fully garbed in early-nineteenth-century style, holding a cane and sitting by a small table on which rest his seal and other personal articles. The cabinet which contains the body is rather like an old-fashioned public telephone box—wooden, with a glass front. For some years now, an artificial, painted wax head has been used, while Bentham's own head is kept in a box in the college cellar. The reason given for this is the "souveniring" of parts of the head that took place some years ago, as a result of which the body was placed in the glass case above an adjacent door. The souveniring continued, and the college removed the head to the basement for safekeeping. Bentham's body is wheeled into the college dining hall twice a year, and toasted—once at the Founders' Dinner and once at the Bentham Memorial Dinner. A copy of the relevant clauses of his will is pinned to the cabinet. If the question whether Jeremy Bentham's body should have been buried has ever been considered, the answer is not apparent. Presumably the gift to the anatomy school is regarded as a sufficient warrant for treating it in this remarkable fashion.

The question whether there is a legal duty to bury a dead body has arisen in Papua, New Guinea, where the influence of English law is still strong and, as we saw earlier, cannibalism involving the bodies of the dead is practised. The particular case was *Regina* v *Noboi-Bosai and Others,* decided by the Supreme Court in 1971. Criminal charges were brought against seven members of the Gabusi people, who inhabit jungle rain-forest country near the Fly River, arising from a payback killing of one, Sumagi, who had been killed by bow and arrow for murdering another villager. By cooking and eating Sumagi's body, the accused (who had nothing to do with his killing) were alleged to have violated a section of the Criminal Code which provided for two years' imprisonment "with hard labour" for "improperly and indecently interfering with a dead human body." The same section provided the same penalty for a person who neglected to perform "any duty ... touching the burial of a human body." The code itself had been prepared by the nineteenth-century English criminal lawyer and judge Sir James Fitzjames Stephen. The judge in the 1971 case remarked on the similarities of this case with the 1884 English trial, *R v Dudley and Stephens* (mentioned earlier in Chapter 5), in which three shipwrecked sailors ate the cabin boy who had been killed by two of them. Those two were charged with murder, but despite extensive research the judge was unable to ascertain what had happened to the third sailor, who, like the accused tribesmen, had partaken of the flesh but had not been party to the killing. His conclusion was that "no proceedings were taken against him." The judge therefore decided that cannibalism was not of itself to be regarded as a crime. The next question was, Did retaining and eating a corpse constitute a neglect or omission to bury it—thus amounting to a crime? The High Court of Australia had said that the continued possession of an unburied corpse would be unlawful only if it was "injurious to the public welfare" and offended public health, public decency, and religion. In the present case, the judge was persuaded that the section of the code dealing with burial was never intended to deal with cannibalism, and that the legislature did not have "in contemplation the banning of a method of disposal of the body, namely by eating, as an alternative to burial or cremation. ... One attempts to apply the standards of the reasonable primitive villager in his proper setting ... not those of the reasonable man on the Clapham omnibus." All the accused were acquitted, demonstrating that the apparently plain words of a law can be inapplicable in circumstances that are unusual enough for a court to conclude that it was never intended to apply to them. The existence of a statutory offence for failing to bury a body was unusual to begin with, but canons of legal interpretation enabled the judge to disregard it.

If Jeremy Bentham and Noboi-Bosai could thread their ways through the legal maze that surrounds dead bodies, the future founders of a bioemporium can be expected to do so too, particularly in the light of laws which allow the removal of human tissues for transplant and therapy. Even though we have been looking at cases and possibilities far out of the ordinary, they emphasize a plain truth. Those who enter areas of changing knowledge and technology to make laws that touch on fundamental aspects of human existence should tread warily and understand that sociological chain reactions are likely. At the same time, they should know that the failure to do anything in the face of inexorable changes may create even worse consequences.

IV.

The very precision of the brain-death definition means that any coma patient whose condition is outside the definition cannot be regarded as dead. When thinking of brain death as legally defined, one could usefully imagine a circle drawn on a piece of paper with the words "coma patients" written across the diameter. One could then visualize a small segment of the circle bearing in smaller letters the words "brain-dead patients." The residue of the circle would carry the words "all other coma patients."

If brain death is defined as "irreversible cessation of all brain function," then it follows that irreversible cessation of only *some* brain function, or even of nearly all brain function, fails to amount to brain death. The person in this condition is alive and is therefore entitled to all the legal rights and privileges of any other living member of the community.

Consider a patient who has brain-stem function but not higher-brain function. The brain is often described as having distinct parts. The higher part, or forebrain, also called the cortex, is the large mass located at the front of the skull. The lower, or more primitive part, located at the back of the skull toward the spinal cord, is called the brain stem, which includes the pons and the medulla oblongata, and the cerebellum. The brain stem is associated with simple bodily functions, and when fully operating is capable of generating heartbeat, blood circulation, respiration, and the working of organs. These can occur even though cortical function has been partly or entirely lost and the patient will forever lack the capacity to be conscious or exercise any of the senses.

In a famous experiment in 1922, two scientists totally removed the higher brains of a number of cats and fed them intravenously, leaving their brain stems untouched. Some of the cats survived with complete bodily function for as long as three weeks. Human beings can be reduced by accidents, or loss of blood circulation for a period, to comparable if not identical conditions of permanent and irreversible unconsciousness, while their bodies remain capable of spontaneous function if given "heroic" medical care and artificial life-support.

Even a layman will be struck by the inclusion of a requirement for repetitive, brief disconnection of life-support machinery under the British criteria of 1976. The doctor must satisfy himself, on more than one occasion, that the patient is incapable of spontaneous respiration. It is rather like saying that he must be 101 percent sure. This suggests that the price of a general acceptance of brain death in the West has possibly been the admission to the category of the brain-dead of a smaller number of hopeless patients than could reasonably be admitted. This restriction, "correct" as it may be, can cause enormous distress for the families of many patients who are excluded. We must despondently concede that the prodigious achievement represented by acceptance of brain death not only has some alarming consequences such as the bioemporium, but is almost certain to cause hardship and tragedy. This is no reason to reject brain-death laws, but merely points up the problems.

The magnitude of the responsibility which must be carried by those who design rules to govern disconnection of life-support machinery and by those who pull the plugs is illustrated by the plight of patients who are totally dependent upon machines for their lives, but whose brains are completely normal. The poliomyelitis victim, or the quadriplegic who requires permanent assisted respiration, must not be accidentally caught by the definition or by the laws which authorize brain-death determination. Such patients are not even inside the imaginary circle called "coma patients," and anybody who deliberately disconnects their life-support machines would be guilty of murder. At one extreme is the comatose, brain-dead patient on a respirator whose body shows many of the attributes of normality but who is in reality a corpse. If no arrangement has been made for the gift of the patient's organs for transplant, the doctor may not only be within the law if he turns off the switch but may be regarded as under a positive moral duty to do so. At the other extreme is the quadriplegic on a respirator; the doctor who deliberately turns off his switch commits murder. In between, in the ranks of the comatose, are those with permanent destruction of part of the brain, who resemble the brain-dead patient in appearance but who will never leave the hospital or open their eyes again. Some are very close to brain death, while others are not, but all of them are inside the "coma circle" while outside the "brain-dead" segment. The doctor who decides that the time has come to disconnect one of these may be about to commit homicide.

Once a statutory definition is adopted, the brain-dead patient becomes in one sense the easiest coma patient to cope with. The real difficulties are with coma patients beyond the reach of the brain-death definition. Probably the best known and most worrying case of this kind is that of Karen Ann Quinlan, the young American woman who fell into a coma in April 1975 when she was twenty-one. At the time of this writing, she is still alive.

Endless interest has been shown the world over in the story of Karen Quinlan. The quantity of words written and spoken in reports, books, scholarly articles, radio and television programs, seminars, discussion groups, churches, and elsewhere is incalculable. It is as if her tragedy embodies all the disturbance we have felt since we were pitchforked by technology a few short years ago into the new world in which death often looks like life, the dividing line between life and death can be decided only by an expert, and dying bodies can be repaired and cured with the body parts of other people.

Karen Quinlan was adopted when only four weeks old by Joseph Quinlan and his wife, Julia. In April 1975, when she was twenty-one, the Quinlans lived in New Jersey, where Joseph Quinlan worked as a drug company section supervisor. They had two other natural children, both younger than Karen. Karen was good at sports, sang in her church choir, and after leaving school at the age of eighteen, took a job as a production worker. She lost her job in 1974 when she was twenty and soon after moved out of home "into a world of casual employment and casual friendships." In 1975, she was described as an occasional marijuana user and "frequent pill popper."

On April 14, 1975, she apparently took some tranquilizers and then went to a bar to celebrate a friend's birthday. She drank gin and tonic and then began to "nod out," according to a friend. It has not been revealed (perhaps nobody ever knew) just what she consumed on that day. Some friends took Karen home and put her to bed, where she passed out. A few moments later, one of her friends noticed that she appeared not to be breathing. He tried mouth-to-mouth resuscitation while an ambulance was called. The ambulance crew managed to get her to breathe again, and rushed her to St. Clare's Hospital in Denville, New Jersey. She has never regained consciousness.

After a few months, Karen's weight had fallen to 60 pounds (27.2 kg). Medical advisers who gave evidence in the first lawsuit which concerned her, in late 1975, used such expressions to describe her condition as "persistent vegetative state" and "grotesque." In the hospital, she had been placed under an intensive care regimen that included three main aids: assisted breathing from a respirator through a tube in her throat, artificial feeding through a tube into her stomach, and the administration of antibiotics to keep infection at bay. Electroencephalography showed some residual function of the brain stem, although there was nothing to indicate that consciousness could be restored. None of the medical experts held out any hope that Karen could ever recover.

Karen Quinlan was not brain-dead.

Her parents were devout Catholics. After their daughter had remained in the same condition for months, and following considerable discussion with their religious advisers, they requested their doctors, and the hospital, to remove the respirator. The doctors refused the request on the ground that they lacked medical authority to terminate life-support systems to a patient who was not brain-dead. They also had moral reservations, and apparently some concern over their legal position, fearing either a malpractice suit or even criminal prosecution under New Jersey law.

The family then applied to the county Superior Court for an order appointing Joseph Quinlan as Karen's guardian, and a further order giving him express power to allow discontinuance of "all extraordinary means of sustaining the vital processes" (Pope Pius XII's statements of 1957 will be recalled here), and an exemption from criminal prosecution for all concerned in the event of such discontinuance.

The legal proceedings attracted worldwide attention. The court was packed every day. On November 10, 1975, Judge Robert Muir handed down his decision. He dismissed the application, holding that there was a compelling community interest in preserving and protecting life without inquiring into or determining the quality or value of that life.

Mr. and Mrs. Quinlan then appealed to the Supreme Court of New Jersey. This court, in effect, reversed Judge Muir's decision and ruled on March 31, 1976, that the parents had the right to seek discontinuance of the respirator. The court said that the community's interest weakens and the individual's right to privacy (i.e. to be allowed to die) grows as the degree of bodily invasion grows and the prospects of recovery recede. It is significant that the court did not itself order that the life-support systems be withdrawn. The judges took the view that this was a medical decision, not a legal one, and should be made upon the request of the parents after consideration by the attending physicians and the ethics committee of the hospital. The medical experts should consider whether there was any reasonable possibility that Karen would ever emerge from her coma "to a cognitive, sapient state" before making their decision.

The decision was made without haste. On May 22, 1976, the respirator was disconnected so that Karen Quinlan might "die." She did not. She began to breathe spontaneously. She was later moved from the hospital to a nursing home where she received the treatment ordinarily given to a chronic patient. Her condition was considered medically to be permanent.

A dreadful irony of the legal proceedings is that nobody, including the courts, appeared to consider the possibility that Karen Quinlan might continue to survive without the respirator. So the same moral, medical, and legal problems applied *after* the fateful day in 1976, this time to the withdrawal of her artificial feeding support and the like.

Many writers have deplored the fact that these questions were ever put to judges, suggesting that the law is not equipped to resolve such a profound problem. The law is often a crude instrument, and certainly seems to be so in a case like this, where a loving family, supported by religious advice and careful medical opinion, is forced to take defensive measures against the laws of homicide. As ridiculous and offensive as this may be, the possibility of prosecution for homicide was strong enough to have been a factor in the Quinlan proceedings. Speaking generally, a person commits the crime of murder if he *intends* to kill another or inflict grievous bodily harm on him and the death of the other person results. The essential ingredient for present purposes is the intention to kill. If you intend the death, the law as written will take no account of your motive no matter how humane. In many

countries, the crime of murder may in certain circumstances be reduced to the less serious crime of manslaughter if the act was not intended to inflict grievous bodily harm or was due to negligence. All of
this means that an act intended deliberately to shorten the life of a
person is a homicide. The consent of the deceased or his family will
not alter this.

Special legal provisions have to be created in some places to deal
justly with the intractable problems of coma, of the helpless aged who
are dying, of the hopelessly defective newborn, and of other human
beings whose existence may be an intolerable burden to themselves,
their families, and the community. If all that a community can offer is
the criminal law of homicide, it will cause a great deal of injustice and
at the same time push heavy burdens and tragedies under the carpet.

The true meaning of "euthanasia" is "an easy or calm death." For over
one hundred years the word has been used to refer to the bringing
about of an easy death, particularly of a person suffering from an incurable and painful disease. New concerns have given rise to new expressions: "active euthanasia" refers to the act of putting someone to
death—for example, by administering a poison or a lethal injection;
"passive euthanasia" and "antidysthanasia" refer to nontreatment or
discontinuance of treatment; and "voluntary euthanasia" refers to the
termination of a patient's life in accordance with his or her wishes.

The law of homicide is not interested in these phrases and does not
recognize their distinctions. Unless statutory attention is given to the
activities described, we will be caught between two unsatisfactory al-
ternatives: the blunt instrument of the criminal courts, and under-the-
counter actions that degrade the participants. The kind of law that can
offer compassionate aid in these dilemmas is California's pioneering
Natural Death Act of September 30, 1976, which established the right
of a person to instruct his medical advisers to withhold or withdraw
life-sustaining procedures in the event of terminal illness or injury.
The Californian lead was quickly followed by the other American
states; ten promptly passed similar laws and in most of the remainder
bills were introduced. A similar bill appeared in March 1977 in the
Ontario legislature in Canada, entitled "An Act Respecting the With-
holding or Withdrawal of Treatment Where Death Is Inevitable." In

1977, the Arkansas legislature directly faced the Quinlan dilemma and provided a legal procedure whereby parents or guardians in a similar case could give suitable directions for withdrawal of artificial life-support.

In this rapidly changing field of medical science, legal obstacles are continually encountered. One source of comfort is the common sense of the normal jury. Despite the law as written, it is hard to believe that a jury today would convict a doctor or a family member who performed an act of so-called euthanasia out of genuine compassion in an extreme instance. In a long line of cases in the United States, and some in England, judges, jurors, and prosecutors have declined to convict or even to bring charges. They range from the deliberate killing of a spouse, or of a child by a parent or vice versa, to prevent further suffering, for example, from cancer, to the termination of life-support by physicians. In other words, the actions of criminal courts in the United States and in some other places tend to belie the written law that equates euthanasia with premeditated murder. Despite this, why should medical practitioners and relatives be jeopardized at all? Their consciences are clear, and they should be given the protection of the law, not put in fear of it.

Somehow Western society must find workable procedures. With an ageing population, decision-making about whether to continue or cease treatment for the helpless, the hopeless, the terminally ill, the permanently unconscious, and the dying is becoming a daily occurrence. One eminent American philosopher and moral theologian, Professor Paul Ramsey, has made encouraging and constructive suggestions that we could follow to advantage. It is not helpful to speak of "murder" or "homicide" in this context, Ramsey claims, because what we are really concerned with is deciding whether a killing is "wrongful" or not. Also, the customary expressions "ordinary treatment" and "extraordinary treatment" in intensive care are meaningless until we invest them with content related to the case in hand. Given rapid technological change, this can be difficult. His opinion is that the most useful result will come from an honest decision by all the doctors and others involved as to the patient's prospects of recovery to a form of living acceptable to the community. If their best conclusion is that no such prospect exists and that continued treatment aimed at producing a cure is nothing but prolongation of dying, then the use of "heroic" or even "standard" measures of curative treatment is no longer indicated.

It would be wrong, says Ramsey—and who would disagree?—if doctors or hospitals could be forced into an endless circuit of treatment because of a demand by a relative or an adviser, aided by obscurity in the law. There must be a power to bring to an end treatment which, is curative in nature—that is, designed to restore a patient to reasonable health—but has no hope of achieving that end. This does not mean that other treatment aimed, for example, at making a dying patient comfortable, should also be discontinued. Maybe it should or maybe it should not. That will depend on the facts of the case.

Both the law and the community should recognize some objective standards of treatment to prevent hospitals or doctors from being forced into procedures based on fear of legal consequences. Conversely, the community needs protection and security against excessive power on the part of the medical profession to decide survival or death without restraint.

Ramsey's view is that in every case a decision should be made at a proper time whether further treatment aimed at a cure is "indicated" or not indicated. If the patient is regarded as dying, the nature of the treatment should then change.

V.

Our new understanding of death and the new utility of the brain-dead are accompanied by many problems. There are also many problems concerning those who could be said to be not dead but nearly so. Do these people have any vulnerability in terms of possible loss of bodily freedom? Obviously, they do. The danger for them is not that their bodies will be treated as "bags of parts" to be dismantled after death for therapy of others, but rather the reverse. Under a kind of paternalistic authoritarianism, they may be prevented from dying when they ought to die or may wish to die, or when they should be humanely allowed to die. There are many reports of the old or the defective being maintained by means of suction devices, tubes, catheters, and other apparatuses stuck into every body orifice and cavity, when there is no chance of recovery to any normal or acceptable way of living. It is for this reason that public pressure has built up across the United States for the exercise of personal choice, voiced by such organizations as the Society for the Right to Die.

As was pointed out in *McFall* v *Shimp,* the living body is frequently subjected to legally based physical compulsion in the normal course of community existence. However, a particular revulsion against the maintenance of life in extreme cases has spread in the wake of the Quinlan decision, and since the California lead in 1976, not only have "natural death" Acts appeared, but allied laws as well, for example, in New Mexico and Arkansas, where it is now lawful to withdraw artificial life-support from terminally ill children with the consent of parents and a court, provided the child does not object. (Arkansas goes even farther, as we have seen, and permits withdrawal, under safeguards, of artificial life-support from adult incompetents, defined so as to include the equivalents of Karen Quinlan.)

This kind of law is aimed at preventing the direct exercise of dominion over a helpless body in a way that many see as inhuman and degrading. One case was reported to a medical journal by a Massachusetts doctor in these words:

> It is true that death is rarely dignified but it is also undignified to die with a urethral Foley catheter connected to a drainage bag, a continuous [intravenous] IV running, a colostomy surrounded with dressings, and irrigation tubes stuck into an abscess cavity around the colostomy, a CVP line, a moisturized oral endotracheal tube attached to a Bennett respirator taped to the face, an oral airway, a feeding nasogastric tube also taped to the face, and all four extremities restrained. This is the way a friend and colleague of mine died. When I went in to greet him two days before he died I could hardly get to the bed because of all the machinery around him. . . . The friend, of course, couldn't speak, and when he lifted his hand, it was checked by a strap. Is it necessary to do this to a human being so his family won't feel guilty about wishing him to have peace at last?

My purpose in this chapter has been to examine the historic change in the perception of human death which has taken place over the past three decades contemporaneously with the rise of modern tissue transplantation. It was the skill of the transplant surgeon, working with skin grafts, that first proved the utility of the human body for therapy: As more tissues of the living became usable, the transplant surgeons began to use the bodies of the newly dead for cornea transplants. Then came viable organ transplants. At the same time, we saw the advent of life-support machines, which showed beyond question that the traditional understanding of death had to be changed, a necessity which was unrelated to transplantation. However, the transplant surgeons quickly saw the immense therapeutic benefits which the brain-dead cadaver offered; it could provide workable spare parts for patients who would die without them. Even without the silent demands of the sick, a redefinition of death would have been essential simply because the classical criteria were no longer accurate for all cases. It was also essential because society could not stand by while the law forced doctors and hospitals to devote the full, complex, and expensive resources of the modern intensive care ward to the twenty-four-hour-a-day care of a corpse. Withdrawal of the artificial maintenance of corpses had to be allowed because the "soft option" of doing nothing visited cruel emotional burdens upon families and medical personnel, with the law of criminal homicide as a background threat to all concerned.

The social and legal recognition of brain death on the one hand, and the development of cadaver transplantation on the other, are a little like the chicken and the egg. Which came first? It seems that transplantation was well under way before the concept of brain death was recognized, but there is an opposite point of view. Certainly society's response to the new utility and value of the human body includes recognition of brain death. It is also certain that brain death as a concept must be permitted to develop in its own right, because of the plight of coma victims who are not brain-dead, and of other helpless old and young patients whose bodies can function only with artificial support.

Transplantation and Free Enterprise: A Commercial Market for Body Goods?

A strong, healthy body is obviously preferable to a feeble or unhealthy one as a source of materials for transplantation, transfusion, therapy, and reproduction. With exceptions for special therapeutic and research projects, the preference is understandable. It follows that organs and tissues of the young and fit will be the most sought-after and the hardest to obtain—in other words, the most valuable. We have already glimpsed in Chapter 1 the strong commercial pressures that can be generated around tissues of both the dead and the living, regardless of social and legal disapproval, when needed by science and medicine. If "value" had the same meaning in relation to the body as it does to articles of commerce, one would expect to find in the West active markets for trade in human tissues, which would be classified according to type, grade, and quality.

Are there markets in bodies and body materials? Whether overt or covert, should markets be encouraged or at least legalized? In recent years, governments have generally answered this last question with a loud no. There is plenty of evidence of official disapproval of this kind of trade and yet, when faced with it, Western society has often behaved erratically and has not practised what it preached.

The 1975 Council of Europe report said that trade in regenerative tissues was "almost universally" illegal, and in nonregenerative tissues illegal in most of its member states, but this statement was inaccurate on two counts. As we have seen, Titmuss cited nine member states which permitted the sale of blood; Britain, for one, has not made commerce in human tissues illegal *as such*. Nonetheless, the prevailing climate of official European opinion is one that flatly opposes commerce. In its model rules of both 1978 and 1979, the Council of Europe recommended outlawing profit-making from activities concerned with removal, use, and transport of human tissues. Specific national prohibitions also exist. The Italian law of 1975 prescribes both imprisonment (six months to three years) and fines (400,000 lire to 2 million lire) as punishment for vendors of body parts. A person acquiring such tissue faces two to five years' imprisonment, a fine of 300,000 to 3 million lire and, if he is a doctor, loss of his license to practise. Outside Europe, trade in human tissues is forbidden in many places, for example, by a 1970 South African law, a 1976 Mexican law, a 1977 Argentinian law, and a 1977 Australian model law.

In order to maintain a balanced view, one must distinguish between tissues. For example, hair and teeth have been traded for centuries in Europe: hair was used by wig makers and teeth were fixed into the jaws of wealthy dental patients as early as the first Elizabethan era. In 1970, *The Times* of London reported the sale of a lock of Lord Byron's hair at Sotheby's auction rooms for £320, while in the same year the Medical Research Council, a government-funded organization, was offering to buy pituitary glands removed from corpses. In 1974, *The Times* ran a letter that complained about the escalating price of skeletons. Legal experts believe that although in general one cannot acquire property rights in human tissues under English common law, there is nothing to prohibit payment being made for them if the parties are acting honestly.

Not to be outdone by human materials, artificial body parts have also achieved the status of trading commodities. In September 1979, it was disclosed that heart pacemakers were being taken from English corpses, "recycled," and implanted in new patients. This practice was uncovered in inquiries that were made following the discovery in Germany of "a macabre, but highly lucrative, trade in pacemakers stolen by doctors from their dead patients." A spokesman for one of the English pacemaker manufacturers was reported as saying: "We know it is not good practice because there is a risk of transmitting infection, but hospitals demand that we recycle pacemakers. It saves them money."

Activities of that kind are either innocuous or socially acceptable,

or sufficiently rare as to be unlikely to provoke widespread public out-
cry or demand for strong legal restraint. But the demands of modern
surgery and research are a different thing altogether. The English
transplant pioneer Professor Roy Calne, speaking at a symposium in
1966, told of cases in which families have sought to buy kidneys from
needy persons for transplantation to sick relatives. We noted in Chapter 1 the practice of wealthy kidney patients taking their live donors
with them to transplant centres. In England, the Secretary of State for
Social Services instituted an official inquiry in September 1976 into
what was described by a mortuary technician as a "wholesale trade" in
the tissues of corpses in a large provincial general hospital. This trade
involved glands, bones, brains, bladders, and other parts needed by
drug companies and researchers. It was carried on without the knowledge of hospital authorities or relatives of the dead patients. In December of the same year, Scotland Yard investigated illicit traffic in human
limbs and organs conducted from the mortuaries of a number of London hospitals. A few months earlier, a leading criminologist from New
York State University acknowledged the likelihood that widespread
black markets in human spare parts would develop as a result of the
new value that they have acquired, and worsening shortages. Despite
the sporadic nature of this information, there is enough of it to conclude that, when considered in the light of never-ending medical and
scientific advances, demands for human body materials, backed by
readiness to pay cash for them, will increase. Confusion is to be expected in nations where it is neither clearly lawful nor clearly unlawful
to sell such materials, especially with important nonregenerative
organs. The parties to a transaction would at the least have to be prepared for adverse public comment if the facts became known, and at
worst for involvement in criminal proceedings. There is always the
possibility of trouble between vendor and purchaser. Persecution and
blackmail of organ recipients by donors have occurred even in cases of
noncommercial tissue donation. For obvious reasons, the true extent
of this commerce is unknown and likely to remain so until laws are
created which reflect society's requirements fairly and make certain
that demand is satisfied.

The most remarkable demonstrations of society's readiness to tolerate trade in human body parts have been provided by the United

States, but even there, approval has not been universal. Americans, however, have evolved a school of thought that strongly and rationally argues the virtues of commerce, presenting it as an alternative both to gift-based systems and to compulsory state power. Analysis of the merits of commerce was encouraged at an early stage by some experts who saw it as inevitable. In 1963, the 1958 Nobel Prize-winning geneticist Joshua Lederberg said at a symposium on the biological future of man that swift medical advances would impose "intolerable economic pressures on transplant sources." Events since then support this opinion. Plainly, the gift-based systems that have predominated in the West have failed to produce the needed quantities of tissues. The European reaction to this failure has been to develop a more peremptory approach. We can consider both the arguments that favour commerce in principle, and the actual experience of it in operation.

Americans who favour commerce are positive about its potential benefits. Since compulsory removal of human tissues will be unacceptable to those who believe in freedom of the individual, they argue that the only way of getting enough tissue is to legalize and encourage sales of body parts. Free commerce is based on individual consent and choice. It is trite economics to observe that any commodity in demand, and at the same time in short supply, will become more valuable. A free-market system assumes that a sufficient number of sellers, enticed by financial reward, would agree to sell their body parts. The supply would be self-regulating. Rising demand would raise the price of tissues in short supply and give more incentive to individuals to sell. Appropriate price levels would be attained, ensuring that organs and other parts sufficient to meet demand would be available.

A number of criticisms can be made on practical and economic grounds. The first is that the mechanics and logistics of body-part supply are very complicated with the most essential organs. With a living seller this problem can be overcome because the time and place of removal (and transplant, if the tissue is to be used in that way) can be arranged to suit the parties. However, when it comes to removal from a dead seller, problems arise. People do not die to order, or in a convenient place. The buyer might never get his goods. The second objection is that great numbers of people would not part with certain body materials at any price. The supply of these parts might prove, in economic jargon, to be inelastic. Or perhaps the only persons who would sell their body parts would be those prepared to give them away in any event—but this objection seems to beg the question and is little more than conjecture. A final objection is that body materials sold in a market may be less healthy than those donated under current systems. This objection rests upon well-known stories concerning blood sales by chronic alcoholics, drug addicts, and down-and-outs who conceal their health defects and other personal information.

Other protests are based on morality, ethics, and social considerations. Many people disapprove of the sale of human body parts on the ground that it is immoral traffic, objectionable in itself and filled with possibilities of abuse, ranging from the exploitation of children and other helpless people to encouragement to murder. As far as dead bodies are concerned, most of us experience a deep emotional reaction to the mere presence of a corpse and are averse to the thought of trade in a dead body or its contents. Under a market system, though, the sale of cadaver organs by next of kin would be permitted. There are not only unsavory aspects to the prospect of a family selling the body of a dead member, but undesirable behaviour can be anticipated. The wishes of the deceased might be ignored because of financial temptation. Even more ghoulish is the thought of a family conference taking place on the subject before the patient has died.

Some critics claim that, under free enterprise, only the poor will sell body parts and only the affluent will be able to buy them, so that the possibility of buying a body or burying one intact could become a luxury available only to the wealthy.

Further opposition to the idea of commerce derives from the general ethical principle that life should be preserved and that an individual should not endanger his own life except for the love of another. This principle has its roots in Judaeo-Christian beliefs that man should never seek his own destruction and that suicide should be condemned. It underlies the common law rules governing modern surgery, body-contact sports, and the voluntary gift of body parts by living donors which were discussed earlier.

With some exceptions, the classical Greeks, the Romans, and modern Western communities have all punished attempted suicide with criminal and religious sanctions. In ancient Athens the hand of a man who attempted suicide was cut off. The Romans, with impeccable logic, decreed forfeiture of a suicide's property when the suicide was committed in order to avoid punishment for an offence that would itself have entailed forfeiture. In medieval England a stake was driven through the heart of a successful suicide and he could not be buried in consecrated ground if he was a Christian. His property was forfeited to the Crown. Modern laws still show vestiges of this attitude which, in the name of preserving life, imposes limits upon the individual's freedom to do with himself as he will. Why should society not recognize a person's right to dispose of his or her life as he or she pleases? A secular answer is that an individual's freedom to do with himself whatever he wishes cannot extend to his own destruction or, for example, to selling himself into slavery, because the offending act will irrevocably destroy the very freedom that makes the transaction possible. Religious disapproval has been voiced by Talmudic scholars and by Plato, who held that it was improper for a person to take his or her own life without divine approval, and by the Christian churches.

Another ethical objection to commerce in human body parts is based on the argument that it also restricts free will and individual autonomy: this is the view that in harming himself or herself, by deliberately undergoing tissue removal, a person may harm society by later becoming sick or enfeebled and a burden upon others. In addition, the person who sells his or her own organs places a surgeon in a difficult position by obliging the performance of what could be a maiming operation.

Turning from principles to practicalities, it is useful to ask what ingredients must be contained in any legal code that regulates this commerce. The first essential is to allocate a legal status to body materials. Then a further series of questions must be answered. These include some which seem normal enough with customary articles of commerce but assume an extraordinary, even fantastic, aspect when asked in relation to body materials:

• Would contracts for the sale of body materials be covered by sale-of-goods laws relating to products liability so that the seller would be liable to pay damages for defects, and for the consequences of breach of normal implied warranties of fitness, and the like? What if the material is diseased or defective or is supplied too early or too late?

• What will the liability of the seller be under laws relating to income tax and capital gains tax? Is the price to be regarded as income or capital gain?

• If a man sells his own body parts for delivery after death, will the proceeds form part of his property and estate for the purposes of death and estate taxes? In a commercial law context, would a free gift of a body part result in a valuation of the part, and the addition of that value to the estate of the donor for the purpose of taxes?

• What kinds of transactions will be permitted? There are various possible contracts, including the sale by a living person of nonvital tissues for immediate payment; the sale by a living person of all or part of the body to be handed over after death but for present payment; the sale by a living person of all or part of the body to be handed over after death with payment to be made to the beneficiaries of the seller; and finally, the sale by next of kin of the body or body parts of a deceased relative.

• Is it really necessary to pay in money? Could not remuneration take the form of free medical care for a period or for life, or the payment of hospital and medical bills, or a kind of "credit" that would give the seller or the seller's family priority in obtaining a transplant should one be needed later? *Time* magazine in October 1973 drew attention to a Kentucky judge who was giving traffic violators the option of paying their fines in blood. The more one thinks about these subjects the more complicated they become, not only because of legal difficulties, but also because of the social and moral dilemmas they pose.

• Trade would presumably be authorized in entire bodies as well as their component tissues. The sale of a dead body could be contemplated, but not the sale of a living person. On the other hand, there seems no reason in principle why the sale of parts from both living and dead persons should not be envisaged, perhaps with reservations concerning the vital organs of the living.

A remarkable illustration of some of these problems comes from *The Book of Lists,* which records a Swedish case of the turn of the century. In 1890, a man made a contract with the Royal Caroline Institute in Stockholm (the medical academy that awards the Nobel Prize for Medicine), according to which he sold his body for anatomical examination after his death. He was in great need of money at the time and was paid immediately. Some years later, he inherited a substantial sum and decided to refund the price and arrange cancellation of the contract. The doctors of the Institute refused, so the seller sued them. He was to learn that, under the law of Sweden, he had signed away his personal autonomy in more ways than one. Not only did the court hold that his body must go to the Institute after death, but he himself was ordered to pay damages. In the intervening years he had had two teeth extracted, and by thus diminishing his body without the Institute's permission he had committed a breach of contract, for which he was penalized.

The question of bestowing legal status upon human tissues is not a new one. The common laws of England and the United States have tackled it on rare occasions, but not definitively. For more than three hundred years, the doctrine has been that "there are no property rights in a corpse." This stems from a pronouncement of Lord Coke, the famous English judge of the early seventeenth century, on the Latin word *cadaver,* commonly used for "corpse." Lord Coke claimed that the word "cadaver" carries the answer to its own question, being, he said, an acronym formed from the opening letters of *caro data vermibus* ("flesh given to the worms"); this was sufficient to conclude that a cadaver could never be treated as "goods" and so could not be the subject of commerce.

Although Coke's opinion was generally accepted as a correct interpretation of the law in common law nations,* it stirred fierce controversy, not so much from its legal conclusions as its allegedly bad Latin. Samuel B. Ruggles, in a New York case in 1856, attacked Coke's scholarship, describing his analysis and opinion as "this verbal conceit, this fantastic and imaginary gift . . . to the worms"; such scholars believed that the true origin of the word "cadaver" is *cadere* ("to fall"), the Romans having used "cadaver" to indicate "something fallen," just as they described the rubble of long-gone cities as *cadavera oppidorum* ("the fallen ruins of cities").

Despite the deficiencies attributed to his Latin, Lord Coke's law was never seriously questioned, although some watering down has been

---

* Common law nations are those such as the United States, Canada, Australia, and New Zealand which have based their legal systems on the common law of England. European nations, speaking generally, have preferred the civil law of imperial Rome as the foundation of their legal systems.

sought in a series of unusual cases. The first occurred around 1750, when a Dr. Handyside was sued in England for an order compelling him to hand over the bodies of "two children that grew together." The tantalizingly incomplete reports of this Siamese-twins case fail to indicate who brought the proceedings, whether the bodies had ever been buried, or how Dr. Handyside came into possession of them; presumably it was the family of the dead twins who sued. In any event, the Chief Justice of Common Pleas, Sir John Willes, declined to order Dr. Handyside to return the bodies, on the grounds that no person can have property in a corpse.

United States law has diverged slightly from the English course, its new path having been marked by a strange case, *In re Beekman Street,* that reached the New York State Supreme Court in 1856. In order to widen this well-known street near New York's City Hall, the city fathers condemned part of an eighteenth-century graveyard in which one Moses Sherwood had been buried for over half a century. Moses Sherwood's daughter objected to this interference with her father's resting place and sued the city for compensation for the expenses of reburying the body. She succeeded, after Mr. Ruggles, a referee in the case, took the opportunity to launch his famous attack on Lord Coke's Latin. Although it seems unlikely that an English court would have awarded damages, the referee did so on the ground of disturbance of the dead, but he was not prepared to hold that any commercial property right existed in a dead body. Later American cases adhered to the English "no property" principle, although some allowed families to recover damages, as in Sherwood's case.

In 1908 the High Court of Australia refused to follow Lord Coke all the way, and held that in some circumstances "a corpse or part thereof is capable of becoming the subject of property." The proceedings concerned the body of a two-headed baby, which had been preserved by a medical practitioner in a bottle filled with spirits and publicly exhibited by a showman named Doodeward, whose father had bought it from the doctor's estate for £36 in 1870. Doodeward was prosecuted and found guilty of outraging public decency. After the trial the police refused to hand back the specimen, so he sued them for its return. He won and regained his "property" on the ground that the body was not an ordinary corpse but had been the subject of skilful work that had changed its nature. At least this decision clarified the legal status of museum collections and assemblies of mummies, shrunken heads, and anatomical specimens. A modern case, in which the old rule appears to have been ignored (or perhaps entirely overlooked), concerns an Englishman who was arrested for a traffic violation in 1974. He was required to give the police a urine sample so that his sobriety could be determined, but after giving it, he poured it down the police station sink before it could be analysed. He was charged with theft of his own body tissue and convicted.

In November 1972, a case was reported in which a transplant surgeon in Athens was charged with "insulting the dead," following the complaint of the family of a deceased hospital patient. The patient, a sixty-four-year-old woman, had been admitted to the hospital in 1969 with a heart ailment but had died suddenly of a stroke. The hospital was given permission by the family to perform a postmortem examination and to amputate a leg, the amputation no doubt having some medical connection with the cause of death. Due to "a legal error," the accused believed that the consent given to amputate the leg also included removal of a kidney. Accordingly, he took one of the deceased's kidneys and transplanted it into a young man who was dying of a kidney disease. When the trial took place three years later, the recipient was fully recovered and leading a normal life. The judge ruled that there had been no "insult to the dead" because no public announcement of the transplant surgery had been made and because the surgeon had charged no fee for it. The charges were dismissed. Although the Council of Europe in 1975 cited Greece as the only member state to permit commerce in human tissues, the lack of a commercial element in this instance had saved the day.

Despite judicial tinkering, or perhaps because of it, the precise legal status of dead bodies and their contents is not clear in either common law or civil law societies.

England seems to be the only country which has stuck fast to the proposition that there can be no property rights in a corpse. More than a century and a quarter after Dr. Handyside's case, Sir James Fitzjames Stephen, in his *History of the Criminal Law* (1883), wrote, "The dead body of a human being is almost the only moveable object known to me which by our law is no one's property, and cannot, so

long at all events as it exists as such, become the property of any one."
In 1976 another legal expert felt compelled to say, "Despite these
murky origins, the rule that there could be no property in a corpse assumed in this country the proportions of an unalterable truth." None
of the other common law or civil law societies of the West, however,
has taken so strict a stance.

Problems concerning bodies and body materials remain unresolved,
generally speaking, so that even apparently straightforward actions
bristle with legal difficulties. Who owns or has the right to possession
of amputed limbs, a removed appendix, gallstones, placentas, and
other bits and pieces? Who is under a duty to dispose of them or to ensure that they do not become dangers to public health? In the United
States these questions are so difficult to answer that courts have assisted parties in sidestepping the "no property" rules by occasionally
permitting them to sue on other grounds. One of the most macabre examples of this occurred in 1975. A patient entered a hospital for patho-
logical tests to determine whether he had cancer of the eye. The
eyeball was removed for examination and clumsily dropped down an
unprotected drain and lost. The Texas Court of Civil Appeals allowed
the patient to recover damages for the negligence, including an
amount for nervous shock. In two other similar cases, however, the
plaintiffs were unsuccessful. The first was brought by a man who suffered from a morbid fear of fire. Some four weeks after the amputation
of one of his legs, he asked what had happened to the severed limb. On
being told that it had been burned in the hospital incinerator, he had
a nervous collapse. The Kentucky Court of Appeals in 1974 rejected
his claim for damages on the grounds that his delay in inquiring about
his leg showed that he was not sufficiently concerned about it. The second case was heard in 1975 in Florida by an appeals court. A mother
was unable to recover damages for the negligent misplacing and loss of
the dead body of her prematurely born baby because she had been
unable to prove that the hospital's actions had been "wanton, wilful,
and malicious."

The United States appears to be unique in the way it has directly faced
the issues raised by commerce in human tissues. It has done so by statute law and court decisions both on a state and federal basis, and the

result is that its laws are neither uniform nor consistent. Nevertheless, a look at the United States experience is certainly of interest.

Some states have enacted legislation that restricts commerce in body parts of the dead. Nevada in 1967, Delaware and Hawaii in 1968, and New York and Oklahoma in 1969 banned payment to a person while alive for his or her body parts after death, but they did not prohibit the sale of the organs by next of kin. Other states dealt with different aspects of commerce, for instance, Mississippi, which in 1969 authorized its citizens to sell their body parts to hospitals, which were given the right to take possession upon death. Breach of the contract necessitated repayment of the money plus interest at 6 percent. Under a 1969 law, Massachusetts prohibited payment to any person for any cadaver organ. Most of these restrictions were abolished when the states successively adopted the Uniform Anatomical Gift Act, since the act was believed to exclude all sales.

There seems to be no express American restriction on the sale of body parts from living persons, and the subject is treated under general principles of law. These principles do much to prevent a sale of vital organs and tissues that would cause gross disfigurement, but have little sway with other tissues, particularly regenerative ones.

The tissue most widely bought and sold in the United States is blood. Payment is also made as a routine matter for urine, skin, and other body fluids such as sweat, saliva, and semen. As long ago as 1970, the ongoing price for blood was $10 to $15 per pint, and for semen somewhere between $15 and $35 per ejaculation. In 1973, a medical research scientist was offering $35 per sample for small pieces of skin. Pituitary glands are frequently bought, and payment is often made to volunteers to undergo experiments. The United States is not the only place in which human body materials are sold, although it seems to be the only Western community in which this is done without causing public outcry.

This is a suitable stage to look at the struggles of legislatures and courts in the United States over the question of whether one vital human tissue—blood—should be classified as a commodity. Since transfusion began, blood in the United States has been acquired both by voluntary donation and by purchase, and blood banks have been owned and conducted by an endless variety of persons and organizations, ranging from unhygienic "mom and pop" shops in slum areas, run solely for private profit, to large community nonprofit associations. In 1970, when Richard Titmuss wrote *The Gift Relationship*, a study of the role of altruism in modern society with particular reference to blood donation, he was able to show that only 7 percent of blood collections in the United States between 1965 and 1967 came from "the voluntary community donor." Over 80 percent came from donors who were either paid or rewarded in some other way. By 1976 such a change had come about in American blood collection that the nonprofit, voluntary American Association of Blood Banks (AABB) found that its members' collections of blood for that year (approaching 3 million pints, equal to 35 percent of the entire national blood collection) contained a mere 8.6 percent from paid donors compared with 18 percent in 1975. In 1975 the AABB and other organizations, under the leadership of the American Medical Association, helped to establish the American Blood Commission (ABC) with the object of "a voluntary, pluralistic and coordinated approach to improving the nation's blood service system." In 1979 a bill was introduced into Congress to incorporate the ABC through a federal charter. These organizations actively support the national blood policy of the federal government, officially announced in July 1973, which set out to ensure adequate supply, attain the highest scientific standards, provide blood for everyone in need irrespective of economic status, and encourage efficient collection, processing, storage, and use.

American courts accepted evidence some years ago that "the risk of transmitting hepatitis is much higher when blood is taken from paid donors." In a 1973 case in Montana, a professor of surgery from Stanford University referred to research studies which showed that blood taken from drug addicts, derelicts, and "skid-row bums" carries a high risk of transmitting hepatitis. He had also conducted tests at the Illinois State Penitentiary which proved that many prisoners there were hepatitis carriers, and he believed that a substantial proportion of the nation's transfused blood came from such sources. (Titmuss estimated that about 5 percent came from prisoners and members of the armed forces in the years 1965–1967.)

Evidence of this kind, coupled with the fact that it is often extremely difficult to detect hepatitis virus in blood, has resulted in a string of lawsuits claiming that blood banks are involved in the sale of goods when they buy blood and dispose of it for a fee, and therefore must carry the same liabilities as any other seller. The rebuttal is that the supply of blood is in reality a "service" and not a "product," and thus no question of products liability or implied warranties of fitness of goods should arise. This counterargument has largely prevailed in the United States for a number of reasons, ranging from those of moral principle (whether it is right that human tissues should be treated as goods) to simple monetary considerations (the cost to the community of allowing product liability).

In the Montana case, an apparently healthy young woman sold a pint of her blood for $5 to a blood bank in Butte in September 1966. The blood was tested according to normal procedures. Three weeks later she came down with hepatitis. The known incubation period for hepatitis made it clear that her blood was infected when she sold it, although nobody knew this at the time. Unfortunately, the blood had already been given to a hospital patient. Despite an immediate warning from the blood bank, it was too late, and the patient contracted hepatitis. He sued the blood bank and the hospital where he received the transfusion for damages. He claimed that the blood was a "product" and that this imposed upon the defendants an automatic liability for defects under products liability law as well as liability for breach of the normal implied warranty relating to the sale of goods, namely, that they are fit for the purpose for which they are intended. The case was tried in 1973, and the judge refused to accept both propositions. The patient who received hepatitis with his transfused blood lost. This decision was consistent with a principle established in 1954 in a New York suit, *Perlmutter* v *Beth David Hospital,* in which the New York State Court of Appeals held that blood transfusion was the rendering of a service and not the provision or sale of a commodity. The *Perlmutter* principle found favour in courts in ten other states during the subsequent fifteen years.

While these cases were being heard in state courts, possibly the most famous proceeding of all, which became known as the Kansas City Case, unfolded at federal level. A contest had developed in Kansas City between commercial blood banks and a nonprofit community blood bank formed on the initiative of local citizens and medical practitioners. By the late 1950s, the community blood bank had a virtual monopoly of blood supply to local hospitals, and the commercial banks decided to complain under the antitrust laws. The Federal

Trade Commission carried out an investigation and in 1962, the community blood bank and its officers were accused of conspiring to restrain interstate trade and commerce in whole human blood. In this charge, blood was equated with goods and commercial products. In 1966, the commission's findings culminated in a decision against the community hospital and its officers on the basis that they were involved in an illegal conspiracy. Appeals were lodged, and the commission's 1966 ruling was set aside by a federal court of appeals in 1969. In the intervening years, numerous bills were introduced into the U.S. Senate in an effort to enact laws to declare that blood is not an article of commerce. They were not passed.

Despite these events, the debate whether the transfusion of blood amounted to the supply of an article or the provision of a service was still going on in 1970. In that year, the Supreme Court of Illinois struck a blow against the *Perlmutter* principle and the proposition that the provision of blood is a service. In *Cunningham* v *MacNeal Memorial Hospital*, the Illinois court rejected *Perlmutter* and the decisions which had followed it in other states and held that products liability law would apply to hospitals and blood banks in Illinois responsible for furnishing hepatitis-infected blood. Soon afterwards, the *Journal of the American Medical Association* published an analysis of the financial ramifications of this decision, which caused a panic among Illinois legislators. Taking Chicago alone, the journal showed that product liability, on a conservative estimate based on the past incidence of transfusion-associated hepatitis (about 750 cases annually), would cost at least $37.5 million a year. The Illinois legislature took one look at the possibility of the mass bankruptcy of its hospitals and, as one commentator dryly put it, "quickly passed a statute nullifying the effect of the *Cunningham* decision and establishing that the use of blood, blood derivatives, human organs, or other human tissue for transfusion or transplantation into the human body constitutes a service and not a sale." By 1975, forty-three states had taken the same step (in fact, some—such as Massachusetts in 1965—had done so before Illinois) and had codified the policy of *Perlmutter* by classifying blood transfusion as a service, not the sale of goods.

Another judgement in this matter was handed down in 1979 by a federal appeals court in the remarkable case of Dorothy Garber.

The defendant was a forty-two-year-old mother of three whose blood was so rare that she was given a commercial contract which many entertainers would envy. Her blood contained an antibody possessed by only two or three other persons in the world, and a laboratory paid her $200 per week for a continuing supply. As an added inducement, she was given $25,000 cash, 1,000 shares of stock in the laboratory corporation itself, and the use of an automobile. The only money which the defendant reported to the government as income were the weekly payments of $200.

When charged with tax evasion in New Orleans, Mrs. Garber raised an unusual defence, claiming that her blood was a "product" and thus entitled to certain special concessions under tax law which were not applicable to service agreements. The circuit court was unimpressed and decided against her in April 1979, on the ground that the sale by a citizen of her blood to a commercial laboratory is not the sale of a product but the sale of a service. She pleaded her case before the U.S. Court of Appeals, which felt the same way as the circuit court and dismissed the appeal, holding that she should have known that all the payments and benefits were income, and should have reported them as such. As a result, Dorothy Garber was convicted of criminal tax evasion.

The preponderant view in the United States, both of courts and legislatures, is that the sale of blood for transfusion is the provision of a service rather than the sale of property, even though there is a market in blood.

In recent years strong moves have been made to eliminate the commercial market in blood, which have considerably reduced the content of bought blood in the national supply. Broad action against commerce began with the proclamation of the national blood policy in 1973. One of the ten aims is that the national blood supply should be obtained from voluntary donors. Further powerful support for reaching this target came on May 15, 1978, with the new Food and Drug Administration regulations that now compel the labeling of all shipments and containers of blood and blood components for transfusion with the words "paid donor" or "volunteer donor," whichever is applicable. Official announcements indicated that the government as well as the courts had accepted the evidence that bought blood in the

United States is more likely to contain hepatitis infection than volunteer blood. There is no medical reason for this, and even the AABB has said that where "a blood-collection program is carefully structured and monitored to assure that its donors do not transmit hepatitis to patients, the fact that its donors are paid has no medical significance." However, facts are facts, and government spokesmen cited official estimates that between ten thousand and thirty thousand people a year had been contracting hepatitis as a result of the transfusion of infected blood, many with fatal consequences. It was also announced that the new regulations were "in keeping with the government policy of moving toward an all-volunteer blood-donor system." The anticommerce tide is rising. In late 1978, the director of the New York Blood Center, which is the largest regional blood-collection agency in the United States, serving over 18 million people, and using over 700,000 pints of blood annually, proclaimed that the goal of the organization is to eliminate commercial blood from New York City.

Another reason for the reaction in the United States against the use of bought blood is the growth of international trade in blood. In recent years "Euroblood" has been imported form Europe in increasing quantities. In 1978, according to the AABB, over 225,000 units were brought in, representing 2 percent of the national supply. There is concern about national vulnerability if this increases, and the grave problems that could be caused by cessation of imports. "It is a national disgrace that we must rely on the life blood of another nation," said the AABB in its statement to the Subcommittee on Health and Scientific Research of the Senate Committee on Labor and Human Resources in June 1979. There are other consequences of importing blood from overseas. The bleeding of poor people in poor countries to supply the requirements of rich countries is fraught with social and political problems. It was reported in 1977, during an uprising in Nicaragua against then dictator Anastasio Somoza, that a plasma-fractionating plant in Managua was burned to the ground by angry mobs. This plant was said to be a major plasma exporter. In 1978 a professor of public administration at New York University claimed that in Bogotá in 1974, a man bled to death trying to support his family. In the same year, the New York press carried stories that international plasma dealers, using Montreal as a transfer point, were operating a flourishing trade in blood obtained from states in the Caribbean. International dealers were also using Third World nations in Africa and elsewhere as sources of commercial blood.

Plasma provides a striking illustration of the potential for abuse of donors. Blood is comprised of a liquid called plasma in which red cells, white cells, and platelets are suspended. The plasma itself contains various chemical substances and proteins vital for maintaining health. As techniques for splitting blood into its components have become more sophisticated, blood plasma is often bought in preference to whole blood because some patients need only plasma or its various components.

By a procedure known as plasmapheresis, the red cells, which carry oxygen to the tissues of the body, can be separated from the plasma during removal and returned immediately to the body of the donor. This means that a donor may give a pint of plasma which can then be "fractionated" for further use, without losing any red cells. Thus, donations can be made at much more frequent intervals than normal, for example every four weeks instead of every six months under a careful community program. However, some commercial schemes envision an even more frequent schedule, and cases have been reported of plasma donors being bled twice a week, at $8 per pint.

Blood supply services in other parts of the world are also opposed to trading in blood. In Australia, all blood for transfusion is obtained on a voluntary basis under the control of the Australian Red Cross Society, which will not become involved in commerce under any circumstances. This extends to a flat refusal to acquire blood or blood fractions from any overseas source, based on the knowledge that in parts of the world the poor and disadvantaged are often exploited for the purpose of making money for others from the sale of their blood. If blood were acquired from overseas sellers, there would be no certain means of checking its source. In Canada there is also an excellent national blood collection and transfusion service, conducted by the Canadian Red Cross Society. This service claims to be "as complete as any in its development and scope," and relies on voluntary donation. At the same time, the Canadian Uniform Human Tissue Gift Act forbids trade in all human tissues with the exception of blood, which is named specifically as a permitted subject of commerce.

There is little doubt that in the West, a money motive behind the disposition of blood and other human tissues is regarded with repugnance, especially if the seller is poor and the recipient rich. Unrestrained growth of trade of this sort might suggest that society has a vested interest in maintaining an impoverished class of citizens to serve as physical risk takers for others. The opportunities for monopolists would be considerable and problems of allocation among recipients could be expected in the event of shortage. This would be intolerable, and the very possibility forces one to conclude that an unrestricted market would not be successful, workable, or acceptable. Government might be obliged to step in and even supervise the market, perhaps diverting its funds from other medical purposes. Even the strongest advocates of commerce in body parts acknowledge that a great deal of legal and economic control would be necessary.

It seems unlikely that widespread commerce in the tissues of human beings, under codes of legal regulation, will come to pass. Still, we can expect that limited commerce will be permitted by law, particularly with regenerative tissues of the living, and there seems to be no reason why donors should not receive proper reimbursement of expenses. Payment to donors for semen, skin, blood, bone marrow, sweat, and urine will ensure that their generosity does not leave them out of pocket. Society should not lose its equilibrium when it comes to trade in tissues. Draconian laws banning every form of commercial transaction may work against the public interest. Legislation in the future seems likely to follow an uneven course in which systems of voluntary consent will be diluted with mixtures of controlled commerce, contracting out, and limited compulsory acquisition. Some communities will tolerate a small amount of commerce, for example in regenerative tissues, and some will allow expansion of the compulsory powers relating to official autopsies. Others will permit exceptions to the principle of consensual giving, but not enough to stir public hostility. Some will go all the way to pure systems of contracting out, and perhaps further, to compulsory removal.

Because of the speed with which events are moving, it is vital that we understand what is at stake.

CHAPTER 8

Special Cases: Artificial Insemination, Test-Tube Babies, and Reproductive Transplants

I.

It has been written that fourteenth-century Arab tribesmen pioneered the technique of artificial insemination by using it as a device to alter the quality and purity of their enemies' horses. Their practice was to deposit secretly in the vaginas of enemy mares cloths soaked with the semen of inferior stallions. Not to be outdone, other writers claim to have discovered traces of this practice among Jews in the second century. Closer to home, biologists accept that the first properly documented account of work in this field was a scientific paper in 1784 by the Italian physiologist Lazzaro Spallanzani that recorded the successful artificial insemination of a bitch. There are also reports that John Hunter, the famous Scottish anatomist and surgeon, attempted artificial insemination in women at the end of the eighteenth century. Spallanzani first thought of freezing human semen in 1776. This was followed by others in 1886 with suggestions for the establishment of banks of frozen semen. The Russians were practising artificial insemination with domestic and farm animals between the two world wars, but it was not until the 1940s and 1950s that artificial insemination of farm animals was undertaken on a massive scale in the West. At the same time, human artificial insemination began to appear as a regular practice.

In terms of the body as property, the use of some tissues essential for reproduction raises the same questions as does the use of other body parts that can be removed and transplanted for therapeutic purposes. However, the ovary and the testicle, both of which have been transplanted, possess characteristics that call for special study. Fetal tissues also have a claim to be treated as special. But semen and ova are in a unique category not so much because of the manner of removal but because of the uses to which they are put. In human artificial insemination (AI), prodigious progress has been made, as it has in *in vitro* fertilization (IVF) and in embryo transplants (ET), which involve the fertilization of a human egg outside the body and the implant of the fertilized egg in a woman's uterus. Problems are posed that have to do with dominion over the body and with possession and identity of human beings, and the materials from which they are created. We are talking now of a new ability to bring human beings into existence not by sexual intercourse but by scientific techniques. The reproductive materials can be isolated, packaged, and stored; a technology has been developed that can create an embryo in the laboratory, arrest its development by freezing, and later thaw it and restart procreation.

The actual number of human artificial inseminations performed from nation to nation is unknown. To some extent, this is because of the uncertain legal status of the procedure, but mainly our ignorance is the result of an absence of regulation and control, which means that it is impossible to ascertain how many people and organizations are carrying out artificial insemination, let alone its total incidence. In December 1972, at a CIBA Foundation Symposium on Law and Ethics of AID and Embryo Transfer held in London, it was disclosed that no nation kept a register recording details of AID (artificial insemination by a donor who is not the recipient's husband). It was also disclosed that scientific estimates of births of AID children in the United States suggested that a total of 100,000 had been born by the year 1957 and that the annual rate was from 5,000 to 10,000 by 1966. By 1975, medicolegal journals were placing American figures at "several hundred thousand persons living today," and the annual rate at around 10,000. In March 1979 *Newsweek* reported that approximately 20,000 women annually were receiving AID in the United States. In 1977, the Royal

College of Obstetricians and Gynaecologists (RCOG) was able to report that it had established communication with twenty-two medical centers in England performing AI either as a private service or as part of the National Health Service. In that year, a total of 2,400 couples had been referred, 1,200 had received treatment, and 731 pregnancies had been achieved. By June 1980, the number of such centers had grown to 38.

The question that immediately arises is: Why? The answer is threefold: first, AI offers a solution to infertility; secondly, there is a rapidly diminishing number of children available for adoption in nations that permit or encourage adoption; and finally, the technique is becoming more and more efficient. (A sidelight on the efficiency of AI as a means of reproduction is available from its use in animal breeding. As long ago as 1964, a Japanese world survey of the use of AI to breed livestock to that time showed a total production of almost 60 million cattle and 50 million sheep. By 1976, AI was the *only* method used in many countries for breeding strains of dairy cattle and other farm animals. Once again, the operative word is "efficiency"—a high rate of pregnancy is achieved, the sperm of the best males is used, the procedure is simple, and breeding rates can be planned.)

What exactly happens in human artificial insemination? The donor's semen is normally obtained by masturbation (it can be obtained surgically) and is deposited by means of an instrument such as a syringe in or near the recipient's cervix. The precise location will vary according to the preference of the medical practitioner. (In Britain, the RCOG found that of the three methods of placing semen, the most favoured was the "intracervical," used in 74 percent of cases, followed by placement in the vaginal vault in 18 percent and inside the uterus in 8 percent. The kind of semen used also varied according to the preference of the practitioner, some favoring fresh, some frozen, and others a mixture of both, while the quantity for each insemination was 0.5 to 1 millilitre with frozen semen, and 1 to 2 millilitres with fresh. More than half the donors were medical and dental students, approximately 11 percent husbands of grateful patients, 11 percent medical staff, 11 percent known to the inseminating doctor, 4 percent laboratory staff, 4 percent other university students, and 4 percent volunteers answering newspaper advertisements.)

The life span of semen once inside the female reproductive tract is short, making the timing of insemination critical. Maximum female fertility occurs at ovulation, when the egg is shed. Although current techniques show a high degree of accuracy, there is no completely certain method in general use for determining the exact time of ovulation. For this reason, some practitioners carry out insemination on a number of successive days in the month, around the time of ovulation; others perform only one insemination per cycle and continue over a number of monthly cycles, although most practitioners prefer not to exceed a year for a course of treatment. Pregnancy rates vary, but many pregnancies occur with the first insemination; many clinics report successful pregnancy in 75 percent of all women treated, with 20 percent becoming pregnant after the first insemination and most of the rest within four months. The overall success rate reported to the RCOG in Britain in 1977 was more than 60 percent from all sources.

Clinical studies have indicated that 10 to 15 percent of all marriages in the United States and Britain are infertile. With about 2.25 million marriages in the United States (the annual figure for 1978) and half a million in Britain (in 1976), the potential demand for AI is great, even accepting that only a small proportion of the infertilities can be cured by AI. (A biologist at the 1972 CIBA Symposium suggested that 10 to 15 percent of the infertilities are due wholly to the husband, which would mean that 10 to 15 percent are theoretical candidates for AID.)

The declining birthrate in Western nations and the diminishing number of children available for adoption are critical factors in the recent growth of human artificial insemination. Except for nations such as France, whose law allows adoption only when parents have no legitimate child of their own, the demand for babies for adoption has so greatly exceeded the numbers available that many couples have felt they had no choice but to try AI. One of the biological papers at the 1972 CIBA Symposium closed with the words: "fewer and fewer babies will be available for adoption. AID therefore fills a real need, and is likely to be used more widely in the future than it is at the present." The chairman of that symposium, Lord Kilbrandon, ended it with words expressing the same sentiment: "Asked 'What do you think of sex?' Marilyn Monroe replied, 'I think sex is here to stay!' I think that AID is here to stay."

This rock-hard fact cannot be gainsaid, as awkward as it may be for certain religions and as unacceptable as it may be to some moralists and ethicists, although it can be, and often is, ignored, avoided, or disapproved.

It is instructive to see how swiftly changes can be wrought (or forced) in official opinion today. A 1948 report of a committee set up by the Archbishop of Canterbury in England stated that artificial insemination was wrong in principle and contrary to Christian beliefs. In 1949 and again in 1951, Pope Pius XII expressed firm papal opposition to all forms of human artificial insemination. The Lutheran Church and the Jewish Orthodox faith take the same view. Yet by 1979, the practice had become so widespread that in Britain the RCOG provided the following written information to prospective recipients:

> Artificial insemination has been practised in this country for many years. Each year several hundred children are born following this procedure, bringing a great deal of happiness to parents. . . . The treatment is straightforward and painless. It will be carried out by a doctor or nurse who will insert a simple instrument into your vagina to place the sperm in the mucus at the neck of the womb. You will probably be asked to rest for a short time after this has been done. . . . However, it is all right to carry on with sexual intercourse as usual. . . .

Clerics of the Church of England are not the only ones who have altered their views on AI. An endocrinologist told a 1976 study group in England that he had successfully given the wife of one clergyman an AID baby and had another eager to start. He had also given AID to a Roman Catholic wife who had received the approval of her priest to proceed. Nor is it difficult to find forceful argument in current Roman Catholic theological writing advocating an alteration in papal opinion and approval of both AIH (artificial insemination with the husband as donor) and AID inside marriage. The problem for moralists and theologians is to accommodate the growing practice and acceptance of AI to the historical Western view of marriage. This tension has been described by G. R. Dunstan, Professor of Moral and Social Theology at King's College, London, who, at the CIBA Foundation Symposium, December 1, 1972, noted that: "Western culture as a whole, in fact, influenced not only by the Judaeo-Christian religion but also by the Graeco-Roman tradition of philosophy and law, has emphasized the nexus between the begetting and conception of children and the shared or common life of the marriage and the family."

Accepting that AID is here to stay, the question is what responses society should make by way of control and regulation. Should anything at all be done? Powerful conservative and reactionary voices the world over have favoured inaction on the ground that guidelines or codes of law would be built on shifting sands and are better not undertaken. This approach is based on ignorance of what is happening in the community, and fails to distinguish between immediate and long-term problems. Many immediate difficulties with AI can be solved by law reform and by positive action that would contribute to community well-being.

Let us try to identify some of the principal problems at present. A hypothetical case will be useful: Charles and Maria had been happily married for five years when they decided to attempt to resort to AID as a means of achieving pregnancy. They were both in their early thirties and had been eager to have a child since the time of their wedding. Despite regular and frequent sexual intercourse, they were unsuccessful, and after three years they had sought medical advice. Tests indicated that Charles had low sperm fertility. They cooperated in a number of schemes to try to achieve conception, some of them requiring a great deal of emotional and physical discipline. (One included the collection of sperm from Charles over a period of time and freezing it for storage after a scientific process of reduction and concentration. Charles was then taught to deposit quantities of this concentrate into Maria's cervix by means of a one-millilitre syringe directed by his finger, immediately before intercourse, as a fortification of his normal ejaculation. Another had involved the doctor himself examining Maria after intercourse and collecting and moving the semen already in her vagina into the cervical canal.) None of these schemes were successful.

At the artificial insemination clinic, Charles and Maria were carefully screened and counselled. They were found to be suitable in all physical and psychological respects for AID parenthood. As it happened, Maria was one of the fortunate 20 percent who became pregnant on her first insemination. She and Charles were delighted. The identity of the donor was not revealed to them, but they knew that the clinic had attempted to provide sperm from a donor who was physically and intellectually similar to Charles. In fact, the donor was a medical student whose personal details were recorded and kept in the confidential files of the clinic. In due course, Maria gave birth to a healthy son, thus providing a satisfying end to six years of childlessness. When he registered the birth, Charles made no reference to the use of artificial insemination and recorded himself as the baby's father.

Unfortunately, Charles did not remain as contented with his household as is normal in AID cases. There is evidence that AID parents are more stable and less liable to divorce than other parents, and that the number of broken marriages among them is substantially lower than in the general population, which may have a lot to do with the strong desire for parenthood and the unusual persistence an infertile couple must evince to produce an AID child. However, this did not apply to Charles, who gradually developed a strong hostility to both Maria and the boy. In another four years they had separated and were preparing for divorce. Charles now said that he was not the child's father and was not willing to pay for maintenance and schooling, claiming this should be entirely Maria's responsibility or the joint responsibility of Maria and the donor, if he could be identified.

The case of Charles and Maria throws into relief the issues precipitated by every AID child concerning family relationships, domestic rights and duties, child welfare, public health, and the accuracy of public records.

- Charles agreed to the use of AID and participated in the entire course of events, and yet the baby is not genetically his. Should he be liable for its maintenance and education? If the answer is yes, should he also have rights of custody?
- Should the genetic father be free of obligations of fatherhood, including maintenance of the child, and conversely, should he be deprived of all custody rights?
- What about inheritance? Should an AID child be entitled to inherit the property of his genetic father or his social father, if either dies without a will? Should either of them be entitled to inherit the child's property in the same circumstances?

In communities that have not abolished the status of illegitimacy, should an AID child be classified as illegitimate?

- Should the performance of AID upon a wife without the consent of her husband be equated with adultery or cruelty or be made unlawful?
- Does the public have an overriding interest in the accuracy of official records, which makes it desirable to place correct information about AI children on the birth register? Should it be a criminal offence to supply deliberately false information about paternity?
- Who should have the right to select donors? What restrictions, if any, should be placed upon them so as to avoid collecting defective or diseased semen, and the too-frequent use of one donor's semen?
- Should donors be paid?
- Should the performance of AI be restricted to medical practitioners?
- Should controls be introduced upon the right to hold, store, and traffic in semen?
- To what extent is secrecy important? Should details of donor, donee, the insemination process itself, and other facts be kept confidential under pain of legal penalty?

In most nations, there are serious legal questions affecting the status and rights of AID children and the practice of insemination itself. A welcome precedent was set by the United States in 1973 when the Commissioners on Uniform State Laws approved the Uniform Parentage Act, which has been recommended for adoption by all states, and in 1974 was approved by the American Bar Association.* The Uniform Parentage Act deals with the legitimacy and family status of the AID child. It says that a consenting husband is to be regarded in law as

* Legislative efforts to help AID children had begun in America as early as 1948, when a bill to recognize and legitimize them was introduced and rejected in New York State. Indiana, Minnesota, Virginia, and Wisconsin also rejected such bills about the same time. By way of contrast, Ohio has rejected a bill which would have made both the medical practitioner and the female patient guilty of criminal offences. However, by 1974, seven states had laws on AID, beginning with Georgia in 1967 and ending with New York in 1974.

the natural father and the donor is not to be regarded as the natural father. The consent must be written and kept in sealed confidential files in the state health department, along with details of the acts of insemination certified by the physician, to be available for inspection only on a court order "for good cause shown."

The Council of Europe has also accepted the desirability of laws on AID. Having excluded from its 1978 model transplant code the "utilization of ova and sperm," the council prepared and published a "Draft Recommendation" on artificial insemination a year later. The draft rules themselves provide primarily for the welfare of the AID child, confine the act of insemination to physicians, require consents from donor, recipient, and husband, and envisage insemination of unmarried as well as married women. The physician is responsible for securing proper consents, checking the health of donors and the quality of semen, and ensuring the secrecy of the parties' identities. Payment for semen and trade in semen is forbidden, but reimbursement of expenses is allowed. The rules specify that an AID child born to a consenting husband and wife shall be their legitimate child and that no legal relationship between the child and the donor will exist. The council accepts that the number of AID children is growing "as a result of difficulties in adoption and of changing social attitudes and technical developments," and expects to see laws in its member states drawn up to "curb any abuses of this practice."

A frequently asked question in England is whether AID without a husband's consent amounts to adultery. One of the few judicial decisions on this matter is a 1958 judgement of the Scottish Court of Session, which held that adultery presupposed some form of sexual intercourse between a man and a woman, and that AI could not be looked upon realistically as sexual intercourse; the court refused to give a nonconsenting husband a divorce for adultery. Other legal decisions had established that sexual intercourse, for these purposes, requires some penetration of the female by the male penis, however slight. (The path of United States case law has not been the same: in 1956, an Illinois appeals court upheld a decision that the AID process amounted to adultery and an AID child was not legitimate.)

One consequence of applying established legal principles to AID in this fashion is the opinion that AID constitutes a criminal conspiracy to produce an illegitimate child. The mentality behind this is one that regards illegitimacy itself as an evil thing, and it shows what strange results derive from applying to AID those laws that were made before it was invented. This is also true of laws of custody, inheritance, and maintenance.

A commentator has the luxury of applauding the benefits offered by new reforms, which may well stretch current public tolerance to the limit, while at the same time arguing that the reforms should go further. Accordingly, with apologies to and applause for the American and European guidelines, I must go on to say that there are serious questions which they failed to address. One is the extent of the AID child's "right to know." To what extent should such a child be told of his or her origins, and be entitled to access to medical records? This question leads to the larger matter of truth in all aspects of AI—not only for the child, but for the community, for science, and for the future.

For many years it was considered unwise to tell an adopted child of his or her origins. This policy has changed, and there is now wide agreement that adopted children have the right to be told by their parents at an early age that they have been adopted, not only because it is impossible to be certain of maintaining secrecy but because it is dishonest to attempt to do so. Research has shown that children are less upset by strange and unpalatable facts than by any form of deception. Much distress has been inflicted upon children who have learned that they are adopted during family quarrels or other crises, or from relatives and friends. With the AID child, the circumstances are comparable but by no means the same. Under present conditions, a great deal of secrecy surrounds the practice of AID because of legal doubts about the donation, the practice itself, and the status of any resulting child. Many doctors either do not make records of donors or, if they do, soon destroy them. One school of thought favours the mixing of semen from a number of donors to protect them and make it impossible to attribute paternity to one person. People have also suggested that a husband's semen be mixed with a donor's, or that the husband and wife have intercourse promptly after an insemination so as to produce similar confusion. Opinions of this kind will persist until the law clearly tells the donor where he stands. If a donor knew where he stood, as he does with the American Uniform Parentage Act, there would be no strong reason to conceal the truth. Most of us have a compulsion to feel a positive identity and wish to know our true genetic background. An AID child, with only partial information, may resort to fantasy and develop emotional problems. Great damage could be done to a person who is told that his or her mother achieved conception from the scientific placement in her cervix of thawed sperm from a donor whose identity can never be recalled or learned. For reasons of this kind, many people believe that AID children should not be told.

Another argument against disclosure is that with conventional adoption, both parents have the same relationship to the child, while with AID, one parent (the social father) is at a disadvantage, which could lead to undesirable repercussions inside and outside the family. Other reasons to keep quiet include these: it will be easier for all concerned; there is always the (remote) possibility that the husband *is* the father, and to rule this out, the child would have to be given information of extreme intimacy, e.g. that the husband was wholly sterile or impotent; the artificial nature of the conception is both difficult to explain and difficult to accept, particularly for a child; revelation of the identity of donors (if this information is in fact available) may discourage donation.

To favour disclosure is, by and large, to emphasize the rights and legitimate expectations of the child, as opposed to the gratification of the couple's desire for parenthood. Good medical reasons also exist for telling the truth. One obstetrician in 1976 drew attention to a family which carried genetic blindness on the husband's side. Their eldest child was born blind. The parents were then advised to proceed by AID, which they did, the wife bearing two children. After careful consideration they decided that the interests of all the children demanded that they should be told the truth. Under present changing conditions, one cannot be dogmatic about telling the truth to all AID children in all circumstances, yet the value of truth, as a matter of principle, is hard to deny, and secrecy and concealment in one area tends to spread to others. It is well known that AID parents frequently supply false information when registering births, but in some places they are *advised* to do so, even by government-supported AID schemes. To supply false information when registering the birth of a child is a criminal offence in most communities, but the possible adverse consequence of writing down the facts can be a powerful incentive to conceal the truth.

It can hardly be doubted that most people would prefer not to be deceived about their ancestry or about the ancestry of others. Falsehood in registration of births or anything else erodes our common interests in truth in public relations and threatens the trust and confidence that bind a society. It could prejudice the development of medical science as well, particularly genetics. Individuals would be deceived, credibility would be reduced, and security diminished if it became accepted professional and official practice deliberately to lie on public records. It is no answer to argue that many children are conceived, as one ethicist put it, out of wedlock already and registered as lawfully begotten. This is not the consequence of an official program. Should records of both genetic and social identity be maintained? The right to the information in the records is a separate question, which can be answered by bringing to bear the same kind of enlightened approach that now characterizes adoption. It follows that deceptive stratagems such as mixing semen (which lacks scientific support) would be discouraged or forbidden. No doubt some people would oppose keeping registers on grounds such as threatened civil liberties, betrayal of anonymity of donors, and the possibility of a computerized Big Brother society, but if justice is to be done, society must strike a fair balance between truth and falsehood.

Another substantial question with AID is "the right to choose"— this refers to the selection of both donors and recipients. A medical practitioner would be unwise to choose a donor or use his semen without carefully checking his medical history and health. Laws relating to professional competence and negligence will apply here, as well as to the performance of the insemination itself. A doctor owes legal duties to the AID child too, and a child born with a disability or defect may have legal rights of action against him.

A curious concern that frequently surfaces in discussion of AID is the worry about "accidental incest"—that is, the possibility that two children produced from the sperm of one donor might meet and unwittingly marry and have children. They would be half-brother and half-sister. There is biological evidence that the progeny of first-cousin marriages show a higher-than-normal rate of malformation and mortality; the children of half-brothers and half-sisters would presumably be under an even greater risk. Most doctors and clinics set a limit of four or five inseminations from one donor, although some impose no limit, claiming that a problem does not in fact arise because the usual donors are medical students, and they are rarely available to a particular department for more than a year at a time! The scientific basis for limiting the number of children from one donor is the desire to reduce both the risks of incestuous marriage and the disproportionate spread of any harmful recessive genes that the donor may carry, but estimates of the likelihood of risk vary widely. A 1960 report in England quoted evidence that if two thousand live AID children were to be born annually in Great Britain, and each donor had been responsible for five children, an accidental incestuous marriage would be unlikely to occur more than once in fifty to a hundred years. (In 1978, an article in the *Australian Mathematical Society Gazette* calculated the probabilities of innocent consanguineous marriages between AID children in Sydney, a city of some three million, at 1 in 40,000 for the offspring of a donor whose sperm produced six AID children, 1 in 60,000 for four children, and 1 in 115,000 for two.)

Should the availability of AID be restricted, for example, to married couples, or should it be the right of any woman to receive AID? At one extreme is the knowledge that in most societies any fertile woman may lawfully become pregnant at any time and by any man she chooses. At the other extreme is an attitude that would deny pregnancy by AID to unmarried women. Ignoring practical problems— limited numbers of medical personnel, limited facilities, and limited quantities of donor sperm—a number of questions still remain. The welfare of the child is of primary importance, which makes it difficult to deny a doctor the right to withhold his services if he wishes to supply them to patients who appear to be well-balanced and mature and have a strong desire to provide a happy home for a child. In early 1978, headline publicity was given in the English press to "the first lesbian AID mother," a woman who was reported to have revealed that she had a six-year-old son conceived by AID and lived in a stable lesbian household with him and another woman. The boy went to school where the facts were known. She said that she found sexual intercourse with men repugnant, and had achieved her pregnancy after being artificially inseminated by a medical practitioner who used his own semen. He had first required her to produce a statement from psychiatrists certifying that she was fit to bring up a child. The publication of this story produced strong adverse reaction, one law journal writer not only expressing his own disapproval of the mother's actions, but suggesting that the medical profession should castigate both the psychiatrists and the inseminating doctor by speaking out "with voices to be heard and marked, roundly condemning such activities as I have described, as contrary to the honourable practice of their profession."

In March 1980, the international press published details of a privately owned sperm bank in California containing semen from Nobel Prize winners. The owner of the bank was seeking women of high intelligence (in the top 2 percent) as recipients. According to a report in *Time* magazine, three American women had already been given semen from Nobel Laureate donors. Observers quickly pointed out that there were no legal controls over this enterprise. It has been known for some time that do-it-yourself artificial insemination kits are available by mail order in the United States. Once again, the question is whether society ought to draw the line, and if so, where.

A comparatively minor topic is payment to donors. It is notable that nations with conservative attitudes toward legislative reform and commerce in human tissues generally are quite prepared to countenance the payment of semen donors. In Britain, the RCOG inquiry of 1977 into the practices of AI clinics discovered that nearly 80 percent of clinics paid their donors, the price varying between £2 and £15 per sample, with a most common fee of £5. Euphemisms such as "reimbursement of expenses" are common. It has been suggested that a normal fee in the United States around that time was $30, while the 1972 CIBA Symposium was informed that a standard payment in Canada was $25. In Australia, AI clinics commonly say that they would be unable to obtain semen if they did not pay for it. It is interesting that the Council of Europe draft rules do not equivocate about payment: "No payment shall be made for donation of semen. However the loss of earnings as well as traveling and other expenses directly caused by the donation may be refunded to the donor. A person or a public or private entity which offers semen for the purpose of artificial insemination shall not do it for profit."

For a variety of reasons this topic cannot be seen in perspective without some conjecture about the future. One is the standard practice of using frozen, stored semen. In 1977, a woman gave birth to her husband's child twenty-one months after his death, following artificial insemination with his frozen semen. This invites analogy with AI in animals and directs attention to selective breeding and eugenics. The spread of AI with farm animals, and its use in many countries as virtually the only means of reproducing certain breeds of cattle, is due to a number of factors. These include efficiency, economy, and sound breeding principles. Animal AI is efficient, and demonstrably effective in its methods of collecting and implanting semen and achieving good birth rates and healthy progeny. The economic advantages are considerable, including the ability to plan a breeding program and the relative cheapness of buying measured quantities of stored semen as opposed to the expense and limitations of buying a stud male or transporting a female to a male. In addition, the semen of proven males, dead and alive, from all parts of the world is available to breeders in local areas anywhere by airfreight.

The technical efficiency of AI for human beings probably has not yet caught up with the same process for animals, but this will change. It will change because the history of human AI has been one of continual technical improvement. A far bigger cause of change, however, will be growing social demand, combined with economic advantages. For economic reasons, the AI trickle with farm animals has become a flood and is now a routine method of reproduction. With human beings, AI is still seen only as a means of overcoming male infertility. However, if it is contemplated not just in this therapeutic role, but as a normal method of reproduction, an utterly different vision of society appears. Let us apply to human AI the first two of the principles which were applied a moment ago to farm animals: efficiency and economy. The subject of selective breeding in human beings is often used as a kind of straw man deliberately set up to be knocked down. The sperm bank of Nobel Prize winners' semen attracted scientific comment that there was no certainty whatever that any resulting children would be of superior intelligence. Critics said that environment, diet, and education are often more important in forming a person than inherited genes, and drew attention to the fact that couples of outstanding intelligence often have produced average and even subnormal children. Comments of this kind, together with an extensive aversion in Western society since World War II to anything that could be labeled as a "master race" plan, have been used as a basis for predicting that human AI will remain a limited practice. However, anyone taking this view may be unable to see the larger picture.

It is easy to anticipate a time shortly when the technique of AI will be a simple, cheap, *and more certain* method of achieving planned pregnancy than any other. We are not talking here of infertile marriages, but normal marriages. AI would eliminate the hit-or-miss fertility of normal sexual intercourse. There would be no need for any change in normal sexual relations. In fact, one could expect an improvement in the enjoyment and the quality of sex because it would be possible for every couple to plan the birth of children with precision and in the meantime to be sexually active and employ contraceptive methods as desired. The advantages of generally available, skilled AI services would be considerable. Every couple could plan the future, giving themselves a much greater sense of present security. They could even increase the likelihood of having children and ensure against the accidental loss of fertility by placing a quantity of the husband's sperm in immediate storage. They could embark on a program of work and money-saving and select future years in which the wife would try to become pregnant and have children. Ovulation could be studied and the best prospects of conception created. When the chosen time came, the AI clinic or the doctor could use the husband's stored or fresh semen. Family-planning in this way appears to be consistent with today's sexual mores and domestic attitudes. The couple intending to use AIH may have much less incentive to seek extramarital sexual relationships. They will be free to indulge their mutual sexual appetites and able to plan their family with the confidence that the prospects of having their own children when they want them will be as high if not higher than those offered by any other method.

This is one of the results that could come from the employment of AI as a routine method of human reproduction. Another, which is not so innocuous, relates to the role of the male. If reproduction by AI became the norm, it would follow that the human male would cease to be socially necessary, at least in the way that he has been in the past. In an

AI world, the individual male may be useful and companionable, but he would not in any practical sense be essential because the comparative numbers of males to females needed to maintain the birthrate would be minute. The human species could easily be reproduced from stored sperm, or from sperm taken from a small number of selected living donors. The social implications of the disappearance of the historic role of the human male are difficult to imagine.

Despite these long-term possibilities, it is unlikely that the continued employment of AI techniques on human beings will provoke serious opposition. By comparison with the creation and birth of a number of test-tube babies since 1978, AI looks almost old-fashioned.

II.

The world's first test-tube baby was born by Caesarean section in England on July 25, 1978, a few days before full term. Louise Brown weighed 5 lb, 12 oz (2.6 kg), and was described by the doctor in charge of the case as a "nice, healthy, normal baby." Her birth attracted enormous attention, comparable to that which followed the first successful heart transplant in 1967, obviously because she was conceived outside the human body, in a small laboratory dish. Louise Brown's mother was incapable of conceiving because her defective fallopian tubes had been surgically removed sometime earlier, but she was fertile and ovulated normally. In November 1977, her medical adviser, Mr. Patrick Steptoe, and his associate, Dr. Robert Edwards, extracted a ripe egg from Mrs. Brown's body and placed it in a dish, where it was mixed with her husband's semen and sustained in special nutrient fluids. After the egg was fertilized, it was transferred to another dish containing nutrient solution in which the process of division into cell clusters began. Between two and three days later, the embryo was implanted in Mrs. Brown's uterus, where it attached itself to the uterine wall and continued to grow into a normal fetus.

These simple statements describe a process of prodigious complexity and uncertain success, which requires medical and scientific skills of a high order. It is plain that we will not witness the overnight appearance of mass-produced test-tube babies. Yet it is also plain that there are irresistible forces at work which will not cease until the procedures are perfected. ("Test-tube baby" has become such a widely used expression that there is little point in dwelling on its misleading aspect; "test-tube embryo" would be more suitable.) One critic has said that it suggests "an almost culinary simplicity" which glosses over scientific procedures of extreme intricacy. The normal medical expressions used in this field are "*in vitro* fertilization" (fertilization in glass), and "embryo transplantation" or "embryo transfer," usually shortened, respectively, to IVF and ET.

Within seven months of the birth of Louise Brown, Steptoe and Edwards had used their method successfully again; a healthy boy was born on January 14, 1979, to Mrs. Grace Montgomery, a thirty-two-year-old cooking teacher in Glasgow, Scotland. In the intervening period, Indian doctors in Calcutta announced the birth of a test-tube baby girl to a thirty-one-year-old married woman. The Indian doctors said that they had used different techniques from the English practitioners, and that after the mother's egg had been fertilized *in vitro,* it had been frozen and stored for fifty-three days before implantation in her uterus.

In early 1979, a test-tube baby clinic was established at Eastern Virginia Medical School in the United States. The techniques of Steptoe and Edwards were to be used, and fees were expected to range from $1,500 to $4,000 per patient.

In early 1980, the expected births of two test-tube babies were announced in Melbourne, Australia, and researchers from Europe, Japan, and the United States visited Melbourne to study the Australian techniques. This emboldened *The Times* of London to predict that "fertilization outside the body may become routine within five years." One of the Melbourne mothers miscarried, but the other, Mrs. Linda Reid, twenty-six, gave birth to the world's fourth IVF child, a healthy girl, on June 23, 1980. By the end of the year Melbourne medical teams were able to demonstrate the dramatic improvement in their IVF pregnancy success rates. In December 1980 it was announced that they had achieved sixteen pregnancies in preceding months, of which seven had spontaneously aborted and nine were continuing normally. These included two sets of twins, which were expected to be born in mid-1981. Considerable progress had been made in the control of ovulation and the technique of implantation of IVF embryos. The pregnancy success rate among women participating in the program was somewhat under 13 percent, with continuing pregnancies (then all between eight and twenty-five weeks) somewhat under 8 percent.

One of the doctors involved in this work, Dr. Alex Lopata, drew attention in the same month to the "wastage" of transferred embryos. In an article in the British journal *Nature* he said that far greater improvements had to be made before IVF techniques would produce high pregnancy rates. In March 1980 the English press reported that Mr. Steptoe and Dr. Edwards hoped to open Britain's first test-tube baby centre, with facilities for offering IVF and ET to some thirty women at a time. By the following September the clinic was an accomplished fact. Called Bourn Hall, in Cambridgeshire, it opened on Sunday, September 28, 1980, with a waiting list of three thousand women prepared to pay £280 for an initial investigation and £1,400 for a ten-day course of treatment. Test-tube baby clinics had also been opened in a number of Britain's leading public hospitals.

The freezing and storage of both fertilized and unfertilized eggs has attracted a great deal of scientific interest. Technically, it has been found simpler to freeze and store semen than eggs; an egg is far bigger than a unit of sperm and contains a variety of tissues which freeze and thaw at different rates, and are thus more susceptible to damage. The problems of successfully freezing and later using a fertilized egg are much less complicated than those for an unfertilized egg. It is easier to freeze and store an embryo than an egg. The use of frozen animal embryos is well-established, as testified by the report in February 1979 in *The Guardian* newspaper of England of a commercial shipment by air from England to New Zealand of fifty frozen cattle embryos for implant into host cows. At the Institute of Animal Physiology in Cambridge, England, healthy lambs and calves have been born after having spent up to three years in suspended animation as frozen embryos prior to implanting in host mothers.*

Scientists working in human cryobiology see no reason why freezing and storage techniques should not be a practical reality in the early 1980s. One English scientist expects to have "some success" by 1982 in

---

* Research teams at the Department of Primary Industry in Queensland, Australia, announced in March 1979 that they had developed methods of producing high-yield test-tube pineapples, papaws, and passion fruits.

freezing an embryo and reimplanting it into a mother. Another expressed the opinion in 1980 that the technique would soon be a reality, as it had been for some time with semen, and that women could expect to bank their frozen eggs and embryos against the risks of future infertility. Quite apart from these purposes, frozen embryos offer increased prospects of success with ET. One of the obstacles to successful implanting is the problem of timing and the need for the mother's uterus to be in the right condition to receive and hold the embryo. If a number of her fertilized eggs could be frozen and stored, more than one could be placed in her uterus during a monthly cycle (or cycles) succeeding the cycle during which the eggs were taken, thus giving greater opportunity for preparation and planning.

The task of ensuring that the mother's body is ready to receive the implanted embryo and hold it raises the question of the host or surrogate mother. There is no technical necessity for the embryo to go into the uterus of the woman who produced the egg. Any woman capable of bearing a child could receive it, carry it, and give birth to the child in due course. The surrogate mother could act as a deputy for the genetic mother, being no more than a vehicle for bearing somebody else's baby, as is common with farm animals. While this is physically feasible, one can readily imagine extreme emotional and mental difficulties with some human surrogate mothers. Such a case came before the High Court in England in 1978. A childless couple approached a young woman and asked her to conceive and bear the husband's child and hand it over to them after birth, for an agreed price. The husband was unwilling to have sexual intercourse with the woman, who was nineteen and had two convictions for prostitution, so she agreed to be artificially inseminated with his sperm and to undertake the pregnancy. The couple had paid another prostitute a finder's fee of about £500 and indicated to the prospective mother that they were willing to turn over almost all their savings if she would surrender the child when born. However, before the pregnancy had run its course, the mother told her clients that she had changed her mind and intended to keep the child. She refused all further offers of money, and after the baby, a boy, was born would not part with it. Three days following the birth, the father and his wife took the child away from the mother against her will. Custody proceedings were begun and the mother won decisions both in the lower court and on appeal, the father being granted limited visitation rights. The appeals court expressed strong disapproval of the entire transaction, labeling the behaviour of the father and his wife as "most selfish and irresponsible." The judge declared that the child's interests were paramount and ordered that he should never be told the circumstances of his conception and birth because of the damage this would do to him.

Despite judicial disapproval, there is no doubt that the era of the host mother has arrived. By 1978, lawyers in California had begun to examine the applicability of the law of contracts, as indicated by the appearance in that year of an article entitled "Contracts to Bear a Child" in an American law journal. One has only to imagine a change of public attitude in order to see the host mother in an entirely different light. There is no legal impediment to the encouragement of any woman, by payment or other reward, to become a host mother, but suppose that encouragement became official and was supported by the community. The role of women is changing so rapidly that it is not difficult to anticipate a time shortly when women may be too heavily involved in important public, commercial, or scientific work to take the time to bear a child, and yet they and their husbands will be able to provide the tissues for an IVF embryo. If healthy young host mothers were offered not only payment but social security, educational facilities, or other indications of public approval, surrogate motherhood could become a normal event. The rapidity with which social attitudes can change from disapproval to unquestioned acceptance is a phenomenon of our times. There is even a curious biblical precedent for surrogate motherhood in the Book of Genesis (30: 1–6):

> And when Rachel saw that she bare Jacob no children, Rachel envied her sister; and said unto Jacob, give me children or else I die.
>
> And Jacob's anger was kindled against Rachel: and he said, Am I in God's stead, who hath withheld from thee the fruit of the womb?
>
> And she said, Behold my maid Bilhah, go in unto her; and she shall bear upon my knees that I may also have children by her.
>
> And she gave him Bilhah her handmaid to wife: and Jacob went in unto her.
>
> And Bilhah conceived, and bare Jacob a son. And Rachel said, God hath judged me, and hath also heard my voice, and hath given me a son. . . .

The biblical case of Rachel and Bilhah was more an example of "symbolic" rather than true surrogate motherhood. Bilhah and Jacob were the parents and the child was conceived naturally, despite the imagery. Equally, the young woman in England who refused to part with her baby was the true biological mother of the child. There is a qualitative difference between such surrogate motherhood and that which results from a test-tube embryo. However, surrogate motherhood by artificial insemination appears to be a growing practice in the United States, according to increasing numbers of press reports. One of these, an article in *Newsweek* in July 1980, described both a medical and a legal practitioner who arranged paid surrogate motherhood as part of their professional practices. Such activity would presumably have been unthinkable without attracting strong social censure even a few years ago.

One puzzling aspect of IVF and ET, at least to non-Americans, has been the absence of United States scientists from the forefront of this field. This has been attributed to some extent to a reported 1975 federal government order whereby the Department of Health, Education and Welfare was barred from providing funds for experimental work unless it was first approved by a national ethics advisory board to be appointed by the Secretary of HEW. According to *Time* magazine in July 1978, the board was not formed until that year. The effect of this delay was to force an unofficial nationwide moratorium on all research having to do with IVF and ET. An additional worry for United States medical practitioners and scientists is the possibility of lawsuits brought for test-tube pregnancies that go wrong. At the very time of Louise Brown's birth, a damages suit for $1.5 million was being heard in New York in which a husband and wife were suing a medical practitioner (among other defendants) who had destroyed a fertilized egg which one of his colleagues had produced in a culture in 1972 and intended to implant in the wife's uterus. The defendant, the head of the department of obstetrics and gynaecology at a university medical centre, said that he had destroyed the specimen because he felt that the skills necessary for success had not been developed at the time and that a monstrosity could have resulted. In addition, the hospital's committee on human experimentation had not approved the action. A jury gave the husband and wife a verdict for $50,000. Irrespective of the possibility that this and similar verdicts could be overruled on appeal, the chances of being dragged into litigation certainly gave American researchers reason to proceed with caution. Malpractice suits by test-tube children against their parents and against the doctors who created them can be easily imagined. The legal status of an egg fertilized *in vitro* must also be considered. Is it alive or not? Can it be discarded and thrown away? A medical practitioner would be ill-advised to embark upon treatment without offering comprehensive advice and obtaining careful documentary clearance from the patient.

Apart from the possibility of proceedings of this kind, which relate to injury, defects, negligence, and breach of professional standards, there seems to be little legal control relating to test-tube babies. The processes are so novel that no nation has any specific laws. Some communities might have public health laws or other rules that could be referred to, but one would not expect to find a law preventing any person from possessing quantities of frozen human semen, frozen eggs, and frozen embryos, or from setting up a commercial business to trade in reproductive tissues. There appears to be no legal power to prevent anybody from fertilizing human eggs in a laboratory or from encouraging or paying a woman to be a surrogate mother, either of an AI child or a test-tube baby made from genetic material unrelated to her, on condition that the child will be handed over at birth. Where should responsibility for the child be placed? More questions crop up as each topic is considered, and they relate to life, death, custody, family relationships, population control, registration, indefinite preservation of embryos, and other vital matters. Who would own all the embryos and all the potential children in an embryo bank where both eggs and semen are held? In a real sense, the scientists who created such an embryo would own it, but what of the child? What if the community decided to take control of all the procedures? Would we then have a state which literally owned its citizens? Would the notion of individual freedom and personal autonomy become moot?

The possibilities are endless, and they all proceed from the fact that the means of reproducing human beings have changed. No longer does mankind beget its progeny only by the direct transmission of reproductive material from one male to one female in the coital act, with the genetic traits of the child determined by those two people. Now we must accept the knowledge that it will soon be a practical possibility to use the semen of one man to father scores of thousands of sons and daughters, and to have the eggs of one woman fertilized outside her body by the semen of numbers of individual men. The resulting embryos could gestate in the bodies of surrogate mothers. As fanciful as these prognostications may seem, they all lie within the range of today's technology. The techniques are already well-established with animals.

There is an abiding desire for children on the part of great numbers of women who cannot now bear their own. Some critics have asked why science should bother to go to so much trouble to bring about conception and birth by artificial means when the world's increasing population is virtually out of control. The answer is that a natural human urge exists to produce offspring, and it is not good enough to point to the children of others as a means of satisfying a woman's desire for her own babies. An estimate produced by Steptoe and Edwards at the 1972 CIBA Symposium was that, conservatively, there were twenty thousand married women in Britain with defective fallopian tubes who could be aided by IVF and ET, and up to seven times that number in the United States. When the proposal to open the Eastern Virginia Medical School's fertility clinic was announced in 1979, a spokesman said that letters were received from as far away as Liberia from women who were desperate to become pregnant and wrote that if necessary they were prepared to sell their homes in order to do so.*

As with human AI, there is no doubt that the techniques of IVF and ET will increase in efficiency and use. Their immediate attraction is that they offer a solution to a basic need of substantial numbers of women and men. They also carry enormous significance for humanity, the implications of which it is too early to assess. It is not easy to say what we should do about it, but it is plain that our interests will not be served by failing to undertake searching inquiry. Among those who are entirely persuaded of the necessity for laws, rules, or guidelines, preferably on an international basis, are the medical experts and scientists who are pioneering this work. They, more than most, can see how great is the potential for abuse.

* The number of women—three thousand—on the Steptoe-Edwards waiting list when they opened their own clinic in September 1980 speaks for itself.

III.

In the human reproductive process, eggs or ova formed inside the ovary travel through the fallopian tubes to the uterus where they await fertilization. During the decade of the 1970s, the world's "first" ovary and fallopian tube transplants were reported at different times in various places. The "first" fallopian tube transplants were announced in Britain in September 1978 and February 1979. In October 1974, reports from West Germany described a fallopian tube transplant which had taken place in Frankfurt using a tube that had been frozen for five months after being obtained from a woman who had undergone a hysterectomy. The recipient was a thirty-one-year-old mother of three who had been sterilized but had later decided that she wished to have more children. Argentinian surgeons were performing transplants of these tissues with considerable expertise in the early 1970s. At the 1972 CIBA Symposium, Dr. Robert Edwards described an operation done in South America in which a woman of twenty-five who had previously had the greater part of her ovaries removed received a whole transplanted ovary after the remnants of her own ovaries had been excised. She later became pregnant. In February 1975, the *British Medical Journal* drew attention to the fact that disease which blocks free passage of ova and spermatozoon through the fallopian tube accounts for about one-quarter of all cases of infertility in women. The journal suggested that transplantation of fallopian tubes is a solution, particularly since, unlike other tissues needed for transplantation, a ready supply of donors exists, namely, healthy tubes from women who have undergone a hysterectomy or sterilization. Immune-rejection problems were said to be considerable but "not insuperable," and it was suggested that the transplants could be synchronized with hysterectomies. These transplants were seen as a preferable alternative to IVF and ET for many women because a successful transplant would avoid complicated laboratory investigations and procedures and would effect a permanent cure.

We saw in Chapter 2 that the first successful testicle transplant was claimed by Beirut in 1972. A similar claim was made in 1977 regarding a thirty-year-old California policeman who was successfully implanted with a testicle from his identical twin in May of that year. Reports state that the case was described some months later to the 1977 annual meeting of the American College of Surgeons; the recipient, who had been born without testicles, had a rising sperm count after the procedure and intended to become a father like his donor brother. According to one of the surgeons involved, two important factors in the decision to carry out this operation were the absence of both immune-rejection problems and genetic differences. He said that the gene makeup of the sperm (and of any child) of the recipient would be determined by the donor's genes, but that this was irrelevant because the parties were identical twins and their gene structures were the same. About a year later, the medical team who performed this transplant announced their intention to carry out a similar one from a father to a son. Dr. Christiaan Barnard described in 1979 how he managed some years earlier to discourage an Italian patient who was strongly pressing him to perform a testicle transplant. The patient brushed aside Barnard's description of the technical difficulties and possibilities of failure. He was taken aback, however, when told that the genetic makeup of offspring would be dictated by the donor and not by him. "You will not be the father of any children, the donor will," said the surgeon. The patient then said he could solve that problem by using one of his father's testicles. "That is worse," replied Barnard, "because your son will then be your brother." The transplant did not take place.

Genetic considerations of this kind have obviously influenced some lawmakers. Those who prepared the Council of Europe model rules of 1978 excluded testicles and ovaries from the general context of tissue removal for transplantation. The exclusion, however, did not extend to fallopian tubes, which are not sites of generation.

The options for action and regulation which are available to society in relation to reproductive tissues vary according to the kind of tissue. We are not talking of all such tissues, but of semen, ova, fetuses and associated material, testicles, ovaries, and fallopian tubes. The uterus and other female sexual parts not mentioned, and the penis, are not envisaged here as sources of therapy or candidates for transplant.

Whether or not the removal and use of testicles and ovaries is permitted for medical purposes will depend on prevailing social attitudes about their proper role in influencing the genetic makeup of children.

Some communities will see no particular significance in implanting organs or tissues of generation (as distinct from other organs and tissues) into one person from another. Other communities will. Apart from the question of genetic makeup, there seems little point in giving special regulatory treatment to ovaries or testicles, and none at all to fallopian tubes, since to do so would be to separate them legally from other nonregenerative tissues.

We saw in Chapters 1 and 2 how dead fetuses and their associated tissues are employed in both research and therapy. Although they are part of the reproductive process, their significance in the present context is different from that of ovaries and the other body parts just mentioned, and from the applications of semen and ova discussed in this chapter. Obviously the use of fetuses merits attention in its own right. The tissues are of a different nature from other body materials and raise separate questions of public policy, morality, ethics, and law. Clearly, religious beliefs will also be influential. No detailed suggestions are made in this book for the legal control of fetal tissue.

As for the collection and use of semen and ova, the most careful consideration is called for. The current stages reached by human AI, IVF, and ET, and the impetus which they have acquired, suggest that every Western nation should make an assessment in the interests of its citizens. An international evaluation based on inquiry into the rate of growth and short- and long-term projections would be even better. With AI, it is plain that prompt law reform could answer urgent questions. There are other less urgent but nonetheless important topics to consider, such as the regulation of sperm banks. Finally, some issues must be addressed over the long run as events unfold. With AI, as with IVF and ET, one of the matters of concern is the lack of regulation of the acquisition, storage, and use of semen and ova. Suddenly a scene has been set where one can imagine these tissues being collected with ease, stored for long, even indefinite periods, united at will in laboratories or workshops, and then put back into storage and suspended animation for as long as desired. The problem is not yet pressing, but it could quickly become so. Few people would contemplate with equanimity the proliferation of storefront embryo banks where passersby could purchase take-home frozen fetuses, or, perhaps more realistically, the storage in a few small refrigeration units of enough reproductive material to create a new nation. The formulation of guidelines, regulations, and laws will require careful thought. As with fetal tissues, this book offers no detailed recommendations because the field is too vast to be dealt with here and needs its own inquiry. The United States Uniformity Commissioners and the Council of Europe have already set encouraging precedents by producing model rules for some of the crucial problems of AI. This initiative should serve as a foundation for further action.

CHAPTER 9

The Ethics of Modern Medical Procedures: Guidelines for Society

I.

What guidelines and principles would you follow if you were charged with preparing up-to-date rules to regulate obtaining human tissues from the dead for therapeutic and community purposes? For the practical application of your rules, imagine that they will apply to a fictitious case.

A patient has just died in the intensive care ward of a large hospital. Her name was Jane, her age twenty-five, and she was married with one child, age two. She and her husband, John, have made wills under which each leaves all assets to the other. Jane died as a result of a simple accident which occurred on a Saturday afternoon following lunch at a friend's apartment. As she and John left, she tripped and fell down a flight of steps, striking her head violently on the corner of a brick wall. Her skull was fractured and by Monday evening, after she had been unconscious for two days, the hospital doctors declared that she was dead by reason of permanent loss of all brain function. Part of the brain itself had been severely damaged by the fall, and the remainder, including the brain stem, had rapidly deteriorated until it too was damaged beyond recall. Jane's parents were present with John at the hospital when she died. She had been connected to a ventilator soon after admission, and this machine kept her heart beating and blood circulating after her body was unable to perform these activities spontaneously. The hospital wished to arrange to transplant Jane's kidneys and corneas. There were waiting recipients. While the final tests were being carried out before death was declared, the medical superintendent spoke to John and to Jane's parents together, and after careful and sympathetic discussion, asked them for permission to remove the necessary body parts. Jane had never given any indication of her wishes one way or the other. John was disposed to agree because he felt sure that Jane would have concurred if she had thought about it and because her death seemed so meaningless otherwise. Jane's parents were not prepared to agree and said that if she died, her body should be removed from the hospital and buried as soon as possible.

An unexpected aspect of cases of this kind is the attention that they focus on the widespread, if not universal, assumption that everyone has the right to direct what shall be done to his or her body not only during life but after death until burial or cremation. It is as though people find it difficult to accept the dissociation of a newly dead body from the group life of which it has formed a part, and wish to retain a kind of autonomy until the body is so fixed in death, or changed in appearance, that it is no longer readily related to the living. The impulse to keep control of our own corpses until interment may be connected to this; for many people, this impulse stems from the age-old belief in resurrection, which requires that the body remain inviolate after death. Nowhere in the West does it seem to have been suggested that a person's control of his or her body ought to cease at the point of death.

Why is it assumed that every person should have the power to direct what shall happen to his or her body until burial? This seems to be another area of human irrationality. If we insist on extending personal autonomy to our corpses, why do we so quietly accept autopsy laws and the gross interference with dead bodies that postmortem examination entails? It is equally puzzling that few communities have bothered to create laws which spell out the obligations of burying the dead. No doubt the laws of hygiene and public health and obscure principles relating to burial grounds and the duties of executors of wills (as in England) could in many cases compel a burial or cremation. If all else failed, we could probably rely on the efficiency of the police and public authorities to collect and dispose of unclaimed corpses. Still, our laws provide no positive direction even though this is a subject on which we encounter passionate insistence on autonomy. Perhaps this very insistence is the reason the laws are so few. People do tend to bury their dead without being told to do so, and in most Western societies, a newly dead corpse is so strongly associated with the once-alive person that nonagreed interference with it affronts humanity.

When it comes to ascertaining the wishes of a deceased person, a number of probabilities must be recognized. The first is that the most suitable donors of body parts for transplant are likely to be people who, like Jane, have never expressed a wish on the subject. The average citizen is unlikely while alive to make an anatomical gift to take effect *postmortem,* and it can be assumed that more often than not the hospital and family will be ignorant of dead patients' wishes. We should therefore acknowledge two other large categories of potential donors: those who gave no consent to the use of their body parts but who voiced no objection, either; and those whose wishes were not known.

Some people strongly believe that the wishes of relatives are paramount when it comes to the removal of tissues from a corpse. According to this view, even if a person makes a gift of body parts, the closest relatives should have a power of veto. Opposed to this is the notion that an express wish of a donor should prevail over any contrary family opinions: most Western legislation reflects this assertion of personal autonomy, for there is clear public benefit in ensuring that a planned donation is not frustrated by relatives or anybody else.

The biggest issue is whether the community has a sufficiently valid claim upon dead bodies to warrant the law's converting it into a public right and legitimizing it for the public benefit. This requires that we determine the limitations to be placed upon the power of an individual and the family to control what will happen to a person's corpse if it should prove suitable for therapeutic use. Resolution of this issue must proceed from knowledge of the medical and administrative means of securing vital organs and tissues. There are hospitals where the most careful consultation takes place with patient or family over the gift of organs, but none occurs in relation to other tissues. It is not uncommon for doctors to remove tissues such as cartilage, bone, tendon, bones of the ear, and even corneas without consent. "You just take out the eye, pop in a wad of cotton wool, and put in a small stitch to keep the lid closed, and nobody is any the wiser," is the sort of statement that perfectly respectable doctors have made to me on more than one occasion.

We can summarize the main choices as follows:

- A legal prohibition may be placed on the removal of any human tissue from a corpse without specific consent from both the deceased and his close relatives.
- The state may be given power to remove any tissue from any dead body irrespective of the wishes of the deceased and the family.
- State power may be softened by allowing the deceased to make an effective veto during his or her lifetime. This is the new European concept, which allows a veto by the deceased but not by the family.
- If there is no indication of attitude by the deceased one way or the other, the family may be given power to consent, to refuse, or to permit removal of organs by indicating non-objection.

Choices may also need to be made on subsidiary topics such as the definition of "family." Who should have an effective voice when it comes to the removal of body parts? Is "family" limited to spouse, adult children, parents, and brothers and sisters? What priorities should exist among family members? What rights, if any, should be given to a "spouse" when the parties were not married, or to divorced or separated spouses?

Other questions include these: Should minors have the power to make effective gifts of their body parts to take effect in the event of death? What procedures should a hospital follow in trying to get in touch with relatives and ascertain their views? Should a hospital have this duty if no relatives are present? Should the coroner have an overriding power to cancel a donation if he has to conduct an autopsy, or should the therapeutic benefit of a proposed tissue gift for transplantation take priority over the coroner's investigation of the cause of death? What penalties or sanctions should be imposed on those who break the laws?

The way in which some Western lawmakers have answered these questions may be seen by examining the current statutes of the United States, France, and England. The United States Uniform Anatomical

Gift Act of 1968 rests upon the requirement of consent.* It is notice-able that the drafter felt it necessary to go to considerable lengths to create a formula for dealing with family involvement in body dona-tion. The formula is elaborate, and both defines "family" and provides an order of priorities.

The French law of December 1976 provides in Article 1: "An organ to be used for therapeutic or scientific purposes may be removed from the cadaver of a person who has not during his lifetime made known his refusal of such procedure. If, however, the cadaver is that of a minor or a mentally defective person, organ removal for transplanta-tion must be authorized by his legal representative." This provision ignores the relatives of the deceased entirely, and avoids all the prob-lems which accompany attempts to embody the consent principle in law.

If we apply separately both the United States and the French laws to the case of Jane, we find that under each one, her parents' wishes will not be legally decisive, and her tissues may lawfully be made available to the hospital without their views being sought. Under the United States Uniform Act, John's wishes will have first priority, and Jane's parents' will have none. Under the French act, the hospital has no obligation to approach any member of Jane's family; by doing so, it does not prejudice its legal powers, and it may remove her organs if it

---

* Subsections 2 (a) and (b) provide:
  (a) Any individual of sound mind and 18 years of age or more may give all or any part of his body for any purpose specified in Section 3, the gift to take effect upon death.
  (b) Any of the following persons, in order of priority stated, when persons in prior classes are not available at the time of death, and in the absence of ac-tual notice of contrary indications by the decedent or actual notice of oppo-sition by a member of the same or a prior class, may give all or any part of the decedent's body for any purpose specified in Section 3:
     (1) the spouse
     (2) an adult son or daughter
     (3) either parent
     (4) adult brother or sister
     (5) a guardian of the person of the decedent at the time of his death
     (6) any other person authorized or under obligation to dispose of the body
  Section 3 provides that the purposes of gifts may be medical or dental educa-tion, research, advancement of medical or dental science, therapy, or transplanta-tion. (See Chapter 3, pp. 71–72.)

wishes. John's agreement will no doubt make the hospital's actions easier from an ethical, but not from a legal, standpoint.

As we saw in Chapter 3, the English Human Tissue Act of 1961 uses concepts that date back to the earliest tissue gift statutes. Its genesis is the Anatomy Act of 1832, and it envisages tissue removal with the consent of the deceased in some cases, and in the absence of objection by the deceased and the family in others. It is deficient on certain critical topics, and public confusion prevails as to its true meaning, but even so, it has been widely copied in other parts of the world and has survived numerous attempts at abolition and reform.

Section 1 of the Act provides:

> If any person, either in writing at any time or orally in the presence of two or more witnesses during his last illness, has expressed a request that his body or any specified part of his body be used after his death for therapeutic purposes or for purposes of medical education or research, the person lawfully in possession of his body after his death may . . . authorize the removal from the body of any part or as the case may be, the specified part. . . .

For cases where the deceased has left no indication of his wishes, the next subsection stipulates:

> . . . the person lawfully in possession of the body of a deceased person may authorize the removal of any part from the body for use for the said purposes if, having made such reasonable inquiry as may be practicable, he has no reason to believe—
> (a) that the deceased had expressed an objection . . . or
> (b) that the surviving spouse or any surviving relative of the deceased objects. . . .

The weaknesses of this statute begin with the fact that it is impossible to know with certainty in all cases who is "the person lawfully in possession" of a dead body. This could result in a hospital or even a government agency deciding to act as "the person lawfully in possession" and peremptorily allowing such tissue removal as it desires. If a cautious hospital decides to comply with the law at all costs, it could find that the statute's ambiguities are so great that frequently it could not ascertain whether it was in breach of the law or not. The Act also confers powers of objection upon "the surviving spouse" and "any surviving relative" but fails to allot an order of priorities among them, and even fails to define "relative." Further, what is "such reasonable inquiry as may be practicable"? How is a hospital to know the extent of its duty to communicate with relatives?

Let us assume that Jane lived in England and the final scene occurred in a major public hospital which had sensibly decided to ignore the possibility of its own superior powers under the Human Tissue Act of 1961, and would remove Jane's tissues only if the family was agreeable. As John is both the husband and the executor of Jane's will, he may properly be regarded as "the person lawfully in possession of" Jane's body. Under the second subsection, quoted above, it appears that Jane's parents will have the power to prevent the removal of her tissue, despite the fact that John as the "spouse" and "the person lawfully in possession of the body" is agreeable, and that the deceased expressed no objection. Both the English and the United States Acts accept the likelihood that dead persons will not have indicated their wishes, and both devote as much attention to allowing relatives to give the body parts as they do to allowing the deceased to do so.

A lawmaker should try to consider the impact of a new law both in the immediate future and in the longer term if foreseeable changes take place (for example, a breakthrough in immunology). A new law in this field could revolutionize official attitudes toward death and take its place alongside all the other legislation that provides some form of control over the bodies of the living and the dead. Looming social issues such as artificial insemination and test-tube babies must not be disregarded. Two subjects can be briefly described to illustrate the responsibilities that these considerations impose. One is of cardinal importance by any standard. The other has little social priority today, but may merit it in the future. The first is brain death, and the second is the freezing and storage of dead bodies with a view to future reanimation.

It is no longer possible to prepare a realistic law on the removal of human tissues and surgical transplantation without deep reflection upon the subject of brain death. Whether the law offers a detailed definition of brain death, simply recognizes it in brief, broad terms, or does neither, there is no avoiding it. Even if a transplant statute confined itself solely to surgical processes and made no mention of brain death, its content and effectiveness would still be dictated by that concept because successful transplants of whole cadaver organs cannot be expected without brain-dead cadavers as donors. A society which permits organ transplantation must allow its medical practitioners to diagnose death by reference to cessation of brain function, whether it does so overtly under the rule of law, or covertly and with self-deception.

Throughout the Western world there are a number of cryonics societies. The members of these societies believe that if a human body is rapidly frozen at the time of death, after the blood has been replaced with a salt solution containing a drug to protect against the damage of freezing, medical science will make it possible in the near future to revive the deceased with his or her personal identity intact. Supporters will also permit their heads to be removed and frozen if for some reason it is not practicable to freeze the entire body, because they also believe that future scientific discoveries will enable the brain to be reanimated and joined to another body. The procedure is lawful by reason of the gift provisions of the Uniform Anatomical Gift Act of the United States. The bodies are kept in long-term storage in capsules or cylinders refrigerated to −196° C. by liquid nitrogen. The center of cryonic preservation is California. It was reported by *Newsweek* in July 1980 that thirty-four bodies had been frozen in this fashion in the United States, most of them in California, including nine held at the premises of Trans Time, Inc., in Berkeley. The idea took root in the mid-1960s, and by 1977, cryonics societies had a membership exceeding one thousand. At that time about $50,000 was needed to pay for permanent preservation, which could be arranged by a special insurance policy.

Cryonics society members have approached governments with serious requests that transplant laws should not interfere with their desires to preserve their own corpses. Not unexpectedly, the subject has attracted wide attention from the press, radio, and television. There is little to justify the creation of specific legal rules to accommodate the wishes of those who favor cryonic preservation, principally because of their minute numbers and the obvious lack of a preferential claim on the time of legislatures. This is not to deny that the cryonicists' beliefs are genuine. It is worth noting that there appears to be nothing in any tissue removal legislation in the United States that would defeat the clearly expressed written wishes of a cryonicist. An objection under a French-style contracting-out law would also veto tissue removal by the state, although there are separate laws in France and other nations of Europe which might overrule the deceased's wishes. As for England, the ambiguities of the Human Tissue Act of 1961 are such that no cryonicist could ever be sure that his wishes would be observed if "the person lawfully in possession" of his body disapproved of them.

Finally, there is the question of breaking the law. No society should enact a statute that accepts the desirability of increased state power over bodies and moves away from the principles of personal autonomy and consent without examining the possibility of abuse. The necessity to prescribe procedures and penalties will to a large extent depend upon the calibre of the society's medical profession, its ethics, and the way in which hospitals are administered, but if there is any danger that hospital behaviour might degrade human dignity, punishment should be considered. On no account should there be any lawful means of treating bodies like carcasses or holding them back from families; they should be handed over in as complete a condition as possible. Public feelings must not be ignored.

If the law is one which requires some form of consent or nonobjection before a hospital may lawfully remove parts from a cadaver, the need to forestall violation may not be so great because all unconsented removals will be illegal. One would not normally expect to find leading hospitals and medical practitioners deliberately embarking on programs of lawbreaking. The English Act has extended this assumption to the limit by failing to include any penalty provisions at all. This has led to more confusion, and to scholarly analysis about the consequences, if any, of contravening its provisions. Those who prepare a consent-based law will be concerned to foster tissue donation and voluntary gifts, but even so, they must recognize that there is a function for penalties and sanctions.

The question of whether or not the human body will be treated as community property is ultimately not one for the lawyers to answer. Lawmakers will play an important role by drawing attention to issues, risks, and dangers, but it is up to the community to decide whether it wishes its corpses to be classified as social resources. Full disclosure of the ramifications of transplantation, intensive care, medical therapy, and the modern miracles of medicine is vital when bodily freedom is to be legally affected in order to accommodate their demands. This is one area in which public opinion should point the way. The law will follow.

II.

If you were asked whether every living citizen ought to be free to give an organ for transplant to a needy recipient, you would very likely say yes. In addition, you might feel that the decision should be the donor's own affair and not the business of the community or the law. However, if you were then asked to prepare guidelines to regulate the removal of body parts from living donors, you would rapidly find that it is not a straightforward matter at all.

Once again a fictitious case can illustrate a number of the problems which a new law should solve. The account which follows contains more medical dilemmas than any normal family is likely to face in a brief period, but each separate crisis could happen to any family at any time.

Roland and Josanne were a married couple in their mid-forties, and the parents of four children. The oldest, Sandy, was nineteen and their only son. The three girls were sixteen-year-old twins, Sue and Jenny, and Liz, aged ten. Roland owned and operated a restaurant. The family had never had any health problems to speak of, but within a year after Sandy's nineteenth birthday, two of the children suffered very serious illnesses. The first was Sandy himself, who woke up one morning with a right eye that was red and discharging pus. During the next three days the eye quickly grew worse, and Sandy felt that part of it was beginning to move forward. At the hospital, doctors explained that the cornea (the transparent part of the eyeball lying over the pupil and iris) had formed into a cone. After some weeks of unsuccessful treatment, Sandy was told that only a cornea transplant would restore his sight. There was no way of predicting when a cornea would become available. He was sent home, heavily bandaged, to wait. While this was happening, Jenny contracted a form of kidney disease which also progressed rapidly. Within four months, it had become necessary to remove both her kidneys in order to arrest a malignant form of blood pressure. She thus became entirely dependent on a dialysis machine, and her doctors advised that she should be given a kidney transplant. Nine months from the first day of Sandy's illness, he was still waiting for a cornea and Jenny had been waiting four months for a kidney.

Roland wanted to volunteer to give one of his corneas to his son, but Sandy refused. The rest of the family, together with their medical advisers, expressed the strong view that his role as head of the family made it necessary for him to remain in good health. Sue had decided very early that she wished Jenny to have one of her kidneys. Because of their close relationship and tissue compatibility, this was not an unexpected decision, and the prospects of success were far higher than with any other donor's kidney. The doctors were not in favor of Josanne donating a cornea to Sandy because she had high blood pressure. This left Liz, who not only adored her big brother, Sandy, but felt left out when she saw her parents and Sue so eager to help. Liz begged to be able to give a cornea to Sandy, and when both he and the family told her that it would not be permitted, she then asked if she could act on her desire to help the sick by enrolling as a blood donor with the blood bank, and as a bone marrow donor with a computerized tissue bank set up by one of the major city hospitals. Accepting, for the completion of this narrative, that both Sandy and Jenny in due course received successful transplants, we now turn to the social questions raised by Liz's wish to enroll as a donor:

- Should there be any legal restriction upon the power of an intelligent, mature person to give his or her body tissues to save another, or should the law stay completely out of it? If the law should play a role, should distinctions be made between different kinds of tissue? Should the law forbid the gift of a vital organ like the heart, or one that will grossly disfigure the donor such as an eye? Should the community foot the bill by providing social service benefits to a citizen who deliberately depletes himself or herself in this way?

- Should a difference be drawn between adults and minors, or should the only standard be the maturity and intelligence of the donor?
- Should limits be placed upon the uses to which tissue donations from living persons may be put?
- Should a legal difference be made between donations of regenerative and nonregenerative tissues, between vital and nonvital parts, and between paired and unpaired organs?
- In our fictitious case, would you allow Roland or Josanne to give a cornea even if they were both healthy and without family responsibilities? Would you allow Sue to give Jenny a kidney, bearing in mind that Sue is a minor of sixteen? Would you allow ten-year-old Liz to give a cornea? If not, would you allow her to give blood and bone marrow, which are both regenerative?

Public debate has taken place on many aspects of tissue removal from living donors. Questions include whether relatives should have any right to participate in making the decision to donate, what kind of health testing for donors should be required, what documentary formalities should be observed, what the standard of medical advice provided to donors should be, and whether anybody should ever be able to give a consent on behalf of a child or a mental incompetent. We have already seen that the judgements in individual cases do not form a recognizable pattern in Western law.

A major concern with live organ donation is the possibility of subsequent emotional disturbance in a donor. In a number of reported cases, donors were content at the time of gift but subsequently displayed changes of personality and developed hostility, even hatred, toward recipients. Depression and fear of future ill health because of the loss have been experienced. There are known cases of regular abuse of a recipient by telephone, letter, and personal visits, plus outright demands for money, and even blackmail. A psychiatric study in 1966 drew attention to unconscious animosity shown toward recipients, and noted that some donors experienced periods of severe depression and concern over the bodily damage that they had sustained, especially a fear that the operation might have resulted in sexual damage.

Preoccupation with sexual matters is not confined to donors. Another study presented evidence that the grafting of a female kidney into a male recipient had produced in the recipient a degree of identity confusion and mental anxiety over his sexual capacity. The new kidney has to be fully integrated, mentally as well as physically. Recipients are often intensely conscious of the new organ lying inside them. Some women have had an initial aversion to sexual intercourse following a transplant, because of fear that male thrusting during the coital act might damage the kidney.

Following donation, some people have become depressed for indeterminate periods. Since a rancorous dependency can develop between recipient and donor if they are known to each other, it has been found necessary to take steps to prevent undue psychological stress. If the prior relationship has been at all ambiguous, this dependency can result in an extremely unhappy and hostile relationship.

This knowledge has prompted many transplant teams to insist on extensive prior psychological testing and absolute privacy of information between unrelated parties, with neither side knowing the identity of the other. The question of privacy is important for the lawmaker, and protective provisions can easily be placed into legislation. This was done with the Australian code and the Council of Europe model code, even though it was realized that the best privacy arrangements can fail if a great deal of publicity attends a particular transplant operation.

Disturbing consequences can also result from the donation of an organ from one family member to another. With adults, as with children, care must be taken to avoid the permanent destruction of good family relationships. On the face of it, nothing could be simpler than a decision by a man to give a kidney to his sister or brother, or by a parent to give one to a child, or vice versa. However, complications can develop if a transplant fails or if the donor subsequently becomes sick or is faced with competing claims, such as the duty to support a wife and children.

In one documented case, a wife described in some detail how a permanent blight was cast on her marriage because of the impact upon her husband's family of his brother's need of a kidney. At the time in question, the late 1960s, the husband's immediate family comprised four adult brothers, one adult sister, and an elderly widowed mother.

Two of the brothers were identical twins and the youngest members of the family. One of the twins, a bachelor, became seriously ill with kidney disease. The other twin was recently married and, with one of the two older brothers who was unmarried, ran a farm. The other older brother was also recently married, to the woman who recounted the story, and the sister was both married and pregnant. The entire family lived in Britain.

The sick twin was put on hospital dialysis in order to survive. His medical advisers suggested that he should seek a transplant and that he should ask his twin to donate a kidney because of the near-certain chance of success between identical twins. He refused to make the request because he did not wish to impose such a burden on his brother and his brother's wife. Nevertheless the healthy twin was quite aware of his potential as a donor, and the buildup of tension in him became so great that his wife succumbed to the pressure, got sick, and had to be given medical treatment.

While this was happening, the elderly mother offered to donate one of her kidneys, but her doctors advised against it because she had recently suffered a heart attack. The married sister was not acceptable because of her pregnancy. Then the healthy twin and his farmer brother decided to undergo tests to ascertain the extent of their compatibility. It was found that their bodies contained a small but distinct level of poisoning caused by farm fertilizers and chemicals. This disqualified them both for tissue donation, which was particularly distressing to the healthy twin.

The result of these events was to focus all family attention upon the fourth brother. The sick twin would not ask for this brother's kidney either, but the emotional pressure placed upon him by the rest of the family and the unusual circumstances was enormous. The kidney patient's doctors then decided that he should be listed as a candidate for a kidney transplant at a leading hospital. After more than a year of waiting, he had risen to the top of the list, but then the transplant unit had to be closed because of an outbreak of infectious disease in the hospital. Soon after this, he contracted a severe illness unrelated to kidney disease and died suddenly.

Ten years later, the family was still fragmented and distressed by these events. The brother who had developed into the candidate for donation was still afflicted by feelings of guilt. His wife said, "My feeling in the whole of this was 'I can't win and my husband can't win because if he has to come and ask my consent I can't refuse it because it would destroy our relationship and if I give it, it will, because I will still resent it' and he could not win either. . . . We still suffer from it." In support of her opinion that the damage to a family can be too great if there is medical or official encouragement of patients to apply pressure to family members, she said, "I feel that sooner or later in life we have to take our chance. Our number is up, and we can't pin half of it on somebody else." When relating the story, she said that she felt more firmly than ever that organ transplants between family members should be prohibited by law if family cohesion is to be retained. This opinion is shared by others, including medical and legal experts who believe that if live donations were legally impossible, intrafamily gifts could not even be considered. The burden of guilt would then be lifted from family members and borne by the community.

Some renal transplant specialists follow a humane, if somewhat deceptive, practice for avoiding the buildup of family pressures upon a chosen "victim." They insist upon lengthy private interviews of all eligible members of the family, irrespective of age. During these interviews, they are particularly anxious to determine whether the chosen person is truly a willing donor. If that person displays any reservation, worry, or hesitation, the doctors refuse to consider him or her as a donor and simply inform the rest of the family that he or she "is not suitable on medical grounds."

It is possible to concede the force of the arguments favoring the banning of intrafamily donations and still conclude that, on balance, personal autonomy should prevail and such donations should be permitted. The increasing number of Western laws which say that adult tissue donation is lawful is a reflection of the view that we should be reasonably free to make decisions about the gift of our own body tissues. In Norway, the 1973 law permits donation, but only for treatment of disease or injury in another, and only if it will cause no danger to the donor's health. In Italy, where kidneys are the only donations allowed, a judge's order must be obtained for all procedures. Mexico permits live donations for therapeutic or teaching purposes provided written consent is given by the donor, but prisoners, mental incompetents, minors, and pregnant women are barred from donation. The 1977 Argentinian law gives permission only for transplantation to relatives, while the French Caillavet Law allows live donations on an unrestricted basis but confines them to "transplantation for therapeutic purposes."

The Australian model code of 1977 is a more comprehensive law which permits live donations of nonregenerative tissue for transplant only. Regenerative tissues may be given for transplant, therapy, or any scientific purpose. In all cases, with the exception of blood transfusion, certain prior conditions must be met, and independent medical advice and a written consent by the donor have to be obtained. The consent must be given in the absence of family and friends to avoid duress. The donor may revoke his consent at any time before any tissue is removed.

The advantages of express legal permission for tissue donation by living adults include the elimination of legal confusion, the creation of clear rules, official endorsement of personal autonomy, and flexibility, which will allow restraints or prohibitions to be placed upon those donations that a community considers undesirable.

Making laws to regulate live donations by persons lacking legal capacity to do so is a far more difficult and controversial task. Some think that those American states that possess judge-made laws permitting removals from minors and mental incompetents have been given compassionate, family-oriented rules for cases which are too hard for most families to decide themselves. Others view that entire legal apparatus with horror.

With minors and mental incompetents, we might first consider regenerative tissue. It may be that a community would be prepared to allow the removal of blood, bone marrow, or skin from a person, despite the fact that he or she is incapable of a fully comprehending consent. Although taking regenerative tissue from a minor involves a breach of physical integrity just as taking nonregenerative tissue does, and also raises the problem of drawing the line between age and maturity, there is a clear qualitative difference. The donor's body will quickly replace the tissue, and the process is relatively simple. Many people are prepared to go along with it for these reasons and, because of the lifesaving capacity of the material, to accept this qualitative difference as a justification for permitting removal from minors. (It will be recalled, however, that this approach did not persuade the court in *McFall* v *Shimp,* where the potential donor was a refusing adult.)

With nonregenerative tissues, there is no meeting point for the two opposing attitudes. Many people passionately believe that no minor should ever be subjected to the removal of nonregenerative body parts, no matter how serious the plight of a potential recipient. One powerful reason is the legal truth that no adult person in the Western world can legally be compelled to donate tissue. To oblige or allow a child to do so when he or she lacks the capacity to make a true decision is considered to be wrong in any circumstances. There are others who believe that the removal of a body part from a minor would be justified in certain cases, and that total prohibition would be unjust because its inflexibility fails to take account of the differences between mature and immature minors. In this view, it is wrong to impose the same blanket prohibition on an intelligent sixteen-year-old as on an intelligent, let alone backward, six-year-old.

Those who would permit donation of nonregenerative tissues by minors tend to advocate safeguards such as independent medical advice, resort to committees of doctors, judges, and psychologists, discussions with the family of the minor, and restriction of recipients to the donor's immediate family, and then only in life-and-death cases. Just as views diverge widely, so do existing laws. Some flatly prohibit the use of minors and incompetents (for example, the 1977 Mexican law), others will allow it on conditions (the French Caillavet Law of 1976 and the 1977 Australian model code), while most fail to deal with the issue at all.

The case for total prohibition of body-part removal from mental incompetents is stronger, but once again it must be conceded that a qualitative approach could be introduced to permit it with some regenerative tissues.

The model rules of the Council of Europe are of particular interest for lawmakers. They permit live donation provided that the donor has received proper advice on its significance and consequences. Privacy and anonymity of both donor and recipient are compulsory except for intrafamily gifts. The donor's consent must be freely given, must be in writing when the tissue is nonregenerative, and also when regenerative if removal presents risks for the donor. There are strict limitations concerning nonregenerative tissues. As a general rule, such tissues are to be given only for transplantation between family members, "except in exceptional cases where there are good chances of success." Prior medical evaluation of risks to both donor and recipient must be made, and the medical profession is ordered to confine all activities to properly equipped and staffed institutions.

The rules also allow removal of tissues from legally incapacitated persons—in other words, from minors and the mentally incompetent. As with the Australian code, there is a general prohibition against taking nonregenerative tissue from the legally incapacitated, but a mechanism is provided which would allow removal in a special case justified for therapeutic or diagnostic reasons. The special case must be confined to a family, and the donor must not only have the capacity to understand what is intended but must consent as well. As for regenerative tissues, the rule allows donation by minors and others without capacity, but only on the conditions that the parent or guardian consents, the donor does not object, there is no risk to the donor, and it is necessary for therapeutic or diagnostic reasons. Thus, it is extremely difficult to take even regenerative tissues from those who lack legal capacity.

The Council of Europe was split on the subject of minors and mental incompetents in the same way as other Western lawmakers:

[A] number of experts expressed an opinion in favor of a total prohibition of the . . . removal of nonregenerative substances [from] . . . legally incapacitated persons, but the majority . . . preferred to bring an exception to it in limited cases. . . . The main reason . . . is the invaluable and irreplaceable nature of substances donated from genetically related persons due to much higher chances of success of transplants effected with these substances. . . . [S]uch a donation by a legally incapacitated person could be in his own interest as well, for instance where [he] has a great interest in the survival of a parent who supports the family and donor. In such a situation the death or deterioration of health of the parent would be much more harmful than ordinary risks of a removal operation.

III.

Another major question for lawmakers is whether there is a need to regulate recipients of human tissue.

A gift cannot be made without a recipient. The very idea of a gift involves a kind of mutuality, and much moral direction is given on the subject, including instruction beginning in childhood that it is better to give than to receive. When it comes to the medical use of human bodies for a community purpose, the idea of a gift is firmly rooted in Western consciousness. In the belief that everyone is entitled to personal autonomy, we became accustomed during the nineteenth and twentieth centuries to the practice of citizens disposing of their bodies to medical schools by including a gift clause in their wills. Social value came to be placed upon the act of giving, so that when modern medicine began to call for ever-increasing supplies of human tissue, the worthiness of donation was accepted. Although the principle of individual freedom will accommodate commerce as readily as gifts, it is interesting that trade in tissue is generally disapproved of. Even in the United States, where tissue removal has been both advocated and practised, note that the name of the statute which regulates it is the Uniform Anatomical *Gift* Act. Some scholars believe that the fundamental nature of the material itself has triggered a primitive human impulse which makes a gift the natural medium. The organization of early societies around gift and reciprocal gift is well known to anthropologists.

The literature and language of transplantation and other medical use of bodies revolve around the principle of giving. The word "donor" is the normal description of the person whose body is the source of tissue, although it obviously applies also to a family member or other authorized person who gives an organ or body part of a patient who has died in a hospital. "Donee" is the complementary expression, but because of its confusing similarity to donor, it has become increasingly common to use the word "recipient" instead.

Not only the person into whose body donated tissue is placed is called a recipient. A recipient may also be a hospital, a medical school, a tissue bank, or a private surgeon or other person to whom a body or part is given. The American Uniform Act probably has the most elaborate provisions of any Western statute on recipients. It considers anatomical gifts to individuals and to institutional recipients, including dental schools, universities, and colleges. All are authorized to accept gifts and given rights and subjected to duties. The recipient's rights to the donated tissues supersede those of all others, except coroners for official autopsies. However, if a recipient is promised tissue by a relative of a deceased person and later discovers that the deceased objected, he is forbidden to accept the gift.

The South African Anatomical Donations and Postmortem Examinations Act of 1970 also provides for gifts to institutional recipients as well as to doctors, dentists, and nominated sick patients.

In Europe, Great Britain, Canada, and Australia there is a noticeable absence of legal regulation of recipients. No attempt is made, for example, in the English Act or the Canadian Uniform Human Tissue Gift Act, to limit the categories or qualifications of donees. It may be that the lawmakers realized that to regulate recipients would be to open a Pandora's box. In an attempt to cover every contingency, you could end up writing an elaborate code which might ultimately prove to be a waste of time. If a law seeks to regulate recipients of body parts, it is obvious that they should be allocated a status. If a dispute were to develop, someone would have to assert a right of ownership or custody, and in the case of theft, the complainant would have to prove a similar right. These considerations are peculiar to recipients. Some writers have suggested that recipients ought to have legal rights against donors of defective tissue, but existing principles of negligence and the law, ethics, and practice rules governing the medical profession already offer many safeguards.

A task which has vexed doctors and other experts is the allocation of priorities between competing recipients. In most communities, this is left to the medical profession or the institutional recipient. Computerized services which collate tissue-typing information have eliminated many of the problems of competing claims. The computer will determine the best match between available tissue and registered potential recipients and give priorities on purely medical grounds, so that available tissues can be sent from one part of a country to another, or overseas if circumstances warrant it. Once again, the need for legal regulation seems questionable. The possibilities of delay and litigation, and the expense of quarrels between claimants, would appear to be sufficient reasons for rejecting it.

Another question for lawmakers is whether to legalize the compulsory implant of tissue. This is the reverse of the problem of competing recipients and typified by the patient who refuses to receive a blood transfusion on religious grounds. Difficulties are compounded when the refusal to allow medical treatment is made by a parent on behalf of a child. In Australia, the only legislation affecting recipients sanctions the administration of blood transfusions without parental consent to save the life of a child. As we saw earlier, courts in the United States are familiar with applications for permission to give compulsory blood transfusions.

Whatever the medical certainties of the benefits of blood transfusion may be, the same certainties do not yet characterize transplantation. There are no good grounds for any form of general compulsory legislation. The impossibility of predicting success in all cases rules out the idea of mandatory implants in minors without parental consent. As for adults, it is difficult either to justify or to imagine the drawing up of public laws that will authorize compulsion, even with blood transfusion, despite the possibility of this happening through court decisions in some American states. With minors and mental incompetents, the combination of parental interest, medical ethics, and hospital standards would produce at least as acceptable a result as detailed legal control. Additionally, no person should be given a transplant or other treatment unless it is intended to benefit him and unless he has freely consented on professional advice. In all normal cases, transplant will be for the recipient's benefit. A problem does arise with the unconscious, emergency patient, but this case would be no different from that of an unconscious patient who needs surgery.

On the assumption that most Western nations have medical professions that exhibit a high standard of competence and ethical behaviour, it is doubtful that good reason exists to pass laws regulating recipients of body-part donations. Nor should donors be compelled to name the recipients of their gifts. If a donor wishes to designate a recipient by name, however, he should be free to do so. The selection of individual recipients does not seem to be an apt subject for legislation, but falls more naturally to the medical profession and the public. Few of the arguments which make legal regulation of donors desirable apply to recipients. Among the latest laws to confirm this are the Caillavet Law and the European model code, neither of which seeks to apply a system of regulation to recipients. It should not be forgotten that because donation has mutuality, there is often an automatic control of recipients when a law imposes controls on donors.

CHAPTER 10

Who Owns Your Body?
Suggestions and Prescriptions

I.

Even in the freest communities, the areas of life in which personal autonomy may function undisturbed are under constant attack. It is not likely that our era will be judged as one which fostered the growth of individual liberty.

While the precise status of human freedom may be a matter of dispute, there is no doubt that television, the press, and other media continually report feelings of impotence on the part of the ordinary citizen. The individual is given daily confirmation by instant information services of his own unimportance and inability to influence the course of events.

For man's future as a social being, there is no more important issue than the just marking of the boundaries of individual liberty. This has special significance for the advanced Western democracies because they still believe that the individual citizen commands respect. Yet even in the West, where the human rights movement has flowered since the end of World War II, the erosion of liberty is constant.

The opening phrases of the United Nations Charter contain these words: "We the peoples of the United Nations determined . . . to reaffirm faith in . . . the dignity and worth of the human person. . . ." In December 1948, the Universal Declaration of Human Rights was adopted by the United Nations General Assembly. Article 3 says: "Everyone has the right to life, liberty and security of the person." The

Western world actively supports the spread and enforcement of the human rights proclaimed in the Universal Declaration. In 1977, three distinguished United States academics, Myres S. McDougal, Harold D. Lasswell, and Lung-chu Chen, carried out a global survey and then listed circumstances for which human rights have been demanded internationally in recent years, and circumstances under which human rights have been denied. They identified 260 separate grievances. Demands have been made for the following bodily rights:

- freedom to accept or reject transplantation or repair
- freedom to accept or reject medical service
- freedom from coerced experimentation
- realization of bodily health and development
- merciful euthanasia

Deprivations of bodily human rights were identified:

- human experimentation without consent
- disputes about genetic engineering, euthanasia, and the like

The demands all reflect important new claims for human rights as they affect the body, and each relates to medical or scientific activities that are expanding fast. However, these activities can move forward only if fueled by a continuous supply of human bodies and body parts. The danger is that too passionate a preaching of their benefits could foreclose inquiry into the nature, or even the existence, of risks. This suggests that the fueling must be monitored to ensure that there is no stampede to make one person's body available to others as a matter of right. A time can be anticipated, if it has not arrived, when large numbers of people will make their living bodies available for the removal of regenerative materials, and donate their dead bodies for treatment of the sick. The law must carefully weigh the interests of the donating individual against the claims of the recipient. If bodies are mortgaged to the community without a full appreciation of the implications of the transaction, the right to redeem them may be lost.

This book has focused attention on a number of extraordinary medical and scientific advances of recent years which carry, along with previously unavailable benefits for the sick, serious implications for personal autonomy. A huge demand has been engendered for a substance of unforeseen therapeutic value—human tissue. Society has been pressured into legalizing and trying to accommodate this appetite. The same influences have also forced an historic change in our understanding of the difference between life and death. Developed nations have responded swiftly to the curative power of transplantation and therapeutic extracts made from body materials. Laws and official programs have burgeoned, facilitating removal of parts from the dead and the living, but demand still outstrips supply. In the field of human reproduction, the future requirements for sperm and ova also show signs of being very large.

In this atmosphere of discovery, society must not lose its head. The claims of the sick and the dying are undeniable, but when the source of therapy is the body of another person, it is time for caution. Common sense should restrain public enthusiasm without stifling it, because developments in the next thirty years will be even more rapid than those of the last thirty. Social attitudes have already undergone a major change, and medical use of body materials is now routine, but it is not clear that society appreciates the ramifications of the new knowledge. Whatever happens, equilibrium should be maintained between the desires of the group and the needs of the individual.

II.

For those who believe in preserving as much freedom as reasonably can be had in the modern world, I have sounded an alert rather than a tocsin, but this could be supplemented by other alarm signals as medical discoveries accelerate and society's interest in the contents of the human body increases. I shall now describe the legal and social regulations I would propose to try to resolve competing interests in a world in which, in yet another contest between the individual and society, our bodies could become the prize.

No purpose would be served by setting down another model code of laws. The foregoing pages have shown that there are already a sufficient number of these to demonstrate the composition and trend of Western legislation.

My own involvement in direct lawmaking concerning the use of human bodies for medical and scientific ends was to head an official task force which produced a model code of laws for the Australian federal government. The code was designed for uniform adoption throughout the Australian federation. It has separate provisions for both the living and the dead, with special attention to minors and mental incompetents, and regulates commerce in human tissues, privacy and confidentiality of medical records, medical liability, autopsies, and anatomy schools. It also legally recognizes the concept of brain death.

The Australian task force was composed of federal judges, legal academics, commissioners of law reform, a legislative draftsman, an immunologist, a philosopher, moral theologians, and fifteen of the nation's leading medical experts including surgeons and physicians specializing in neurosurgery, neurology, intensive care, anaesthetics, anatomy, and the transplantation of hearts, kidneys, and other tissues. Extensive use was made of public hearings, and of the press, radio, and television. The inquiry covered transplantation, general therapy of the sick, research, experimentation, autopsy, and anatomy studies, all of which are claimants for rights of access to the body. However, in this book I am envisaging something larger, and that is an unavoidable future multiplication of official attempts to exercise direct dominion over the human body solely because of medical and scientific progress.

These attempts will necessarily be of a peculiarly physical kind, interfering with bodily integrity. Surgical transplantation obviously involves cutting open the source body and removing some of its contents. So does the preparation of therapeutic substances from human tissues and the conduct of autopsies, anatomy study, and many research projects. Even with regenerative tissues, concerning which society is more likely to tolerate the idea of compulsion, the process of removal is not always simple or painless.

New medical techniques reach out for physical dominion over the body in new ways. Medicine's ability to keep alive human beings who in the past could never have survived increases continually. The process of dying can be agonizingly drawn out with the very aged by means of all manner of machines, drips, suction devices, and artificial apparatuses. The process of living can be, too, with newborn babies who have incurable mental and physical defects. Such treatment can be administered to the permanently unconscious, including those whose brains are partly, or largely, if not entirely, destroyed. Almost insoluble problems are raised for society by the doctor who, for long periods, artificially supports the lives of dying, semiconscious, senile, incompetent people who would have been appalled at the thought of such treatment when they were younger and had normal mental faculties. Many permanently unconscious patients cause similar dilemmas. They are not dead by the newly accepted concept of brain death because they have not lost all brain function. They are therefore alive by definition, and the laws of homicide stare at anybody who wishes to end their suffering. The Karen Ann Quinlans of this world can never recover, and their bodies, together with the other nonviable bodies described above, present difficulties which make many philosophers, politicians, doctors, and lawmakers run for cover. Somehow or other society must grapple with these issues because they will not go away. If anything, they will become more common.

Another example of the assertion of direct dominion over the human body is the creed of the antiabortionist. The abortion debate stirs deep moral and religious passions and continues to rage. At the very heart of it is the issue of personal autonomy. Whichever way the argument is resolved for a particular community, the power to control the pregnant woman's body is at stake. If abortion is forbidden by law, women lose personal autonomy and are unable to deal with important functions of their own bodies without the risk of committing a criminal offence. If abortion is permitted by law, a woman will have the power to determine what will happen to her body and the embryo in it.

The physical dominion we are talking about here is principally that which controls the removal of body materials for subsequent use. This kind of dominion extends to semen, ova, fetuses, ovaries, fallopian tubes, and testicles, as well as to any other tissues. However, growing control over the reproductive process brings up other considerations. The implications of artificial insemination and test-tube babies are of a separate kind from those we are now discussing because they refer to the status of people who will be brought into existence by unusual and unprecedented methods, and to the role of those who are involved in their creation. Therefore I will make a value judgement here and exclude semen, ova, and fetal tissue from the operation of my law for the reason that they raise discrete problems of major importance.

Artificial insemination, the fertilization of a human egg outside the human body, the implanting of a fertilized egg in a woman, and all other uses to which eggs and donated semen can be put for reproductive purposes should be the focus of separate guidelines. So should the use of fetuses, because this subject is inextricably bound up with methods of abortion and raises substantial questions of public policy, morality, and law. The same reasoning resulted in the exclusion of "fetal tissue, spermatozoa [and] ova" from the Australian model code and of "the transfer of embryos, the removal and transplantation of testicles and ovaries and the utilization of ova and sperm" from the Council of Europe model code. My law would apply to ovaries, fallopian tubes, and testicles because they do not, in my view, truly raise different problems from those occasioned by the removal of tissues generally.

If the body is studied as an object in which rights may be claimed or acquired by others, links can be discerned between events which otherwise have no apparent connection. The vulnerability of the body to treatment as property can readily be seen by using this classification and temporarily ignoring noncorporeal attributes such as spirit, soul, and personality. It enables an observer to recognize the chasm between the motivation of a slave owner and the motivation of a transplant surgeon eager to secure supplies of body parts, and yet to acknowledge a relationship between them based on the legal treatment of slaves as chattels and the potential legal recognition of claims by the sick to the contents of healthy human bodies. Present and future threats to the integrity of the human body and to individual liberty can be illustrated by the treatment of the body as a kind of property with the support of the legal system.

If I were to prepare a new law aimed at allowing the human body and its contents to be placed in the service of the community and the sick for the purposes which we have been discussing, I would accept a series of distinctions and facts:

- Separate treatment must be given to the bodies of the living and the dead.

- With the living, differences must be discerned between adults, minors, mental incompetents, persons with other disabilities, and persons of varying maturity.
- There is significance in the distinction between regenerative and nonregenerative tissues, paired and nonpaired organs, and vital and nonvital tissues.
- The closer the blood relationship between donor and recipient in live transplantation, the greater will be the chances of success.
- The role of the family of a donor must be considered.
- The dead hospital patient has a particular importance in transplantation, as does the concept of death by reference to cessation of brain function.
- The relationship between the law to be created and other laws which directly affect the human body must be taken into account. These will include legislation relating to coroners and those who conduct official autopsies.

There will be other factors to keep in mind. One is the need to guard against creating unnecessary legislation. The American story of a disease called phenylketonuria is a good lesson in kind. Commonly referred to as PKU, this is an hereditary metabolic disorder which may cause severe mental retardation. Its incidence is so low in the United States that even among children in institutions for the mentally retarded, it affects less than 1 percent. In the early 1960s, popular scientific writing, publicity, and a public expectation that medicine could cure or prevent any disease if given the money and the opportunity led to widespread pressure for universal, mandatory laws to test all newborn babies for PKU. Between 1963 and 1976, forty-one states passed such statutes. By 1976, with mounting evidence that the legislation was premature and that scientific understanding of PKU was incomplete, wiser counsel began to prevail. The consequences of incorrect diagnosis were causing concern because of the severe damage that can be done to a normal child who is compulsorily treated for a mental disorder which he or she does not have. Attention was drawn to infants' diseases of wider prevalence, such as heart complaints which could have received the same money, time, and attention with more benefit to the community. The provision of general diagnostic facilities for the

99.1 percent of children in institutions who did not have PKU began to seem far more sensible than the specialized arrangements made to identify the 0.89 percent who did. Unnecessary lawmaking of this kind can be as undesirable as failing to act at all.

My most important task would be to formulate coherent underlying principles so that the law will be a consistent and credible public statement. Three questions must be answered satisfactorily before a law which facilitates the removal of organs and tissues from both the living and the dead will be justified. The first is whether the use of human tissue for transplantation, therapy, and other purposes is desirable, in the community interest, and to be encouraged. This book has already demonstrated that society's answer is yes. The second question is whether there is an adequate supply of tissue, and the answer is, for reasons also given earlier, no. The final question asks what steps should be taken to increase the supply. The chapters dealing with voluntary systems, compulsory systems, and a commercial market present the possible solutions.

The shape and character of the law will depend on the decision whether tissues (and bodies) should come from a "giving" source or a "taking" source. Should the philosophy of consent-based donation, which relies solely upon individual generosity and leaves the initiative to the citizen, rule the supply system? Or should the community stake its claim by allowing bodies to be used as needed and place the onus of refusal upon the individual citizen, so that he or she must contract or opt out?

I would base all my decisions upon the belief that the individual is not primarily some kind of social debtor whose obligations to the community outweigh, or do no more than balance, his rights, powers, and privileges. Society and its laws should aim to promote personal autonomy and individual liberty. Accordingly, my law in relation to tissue removal from living persons would be expressed positively, not negatively, and would allow donations of body parts by adults of sound mind, provided that it is done on the basis of free and informed consent.

I would, however, restrict the use of donated nonregenerative body parts to the purposes of transplantation. Transplantation has merit because of its lifesaving nature and prospects of success. The use of nonregenerative tissues for other purposes, for example, research, cannot offer the same direct value, and the interests of both the individual and the public are better served by not allowing such donations by the living. My law would contain no provisions restraining the donation of vital single organs or of body parts for transplant purposes which may result in undue maiming. My opinion is that such donations should not occur, but I would leave their control to common sense and medical ethics rather than fill the statute with prohibitions of all foreseeable undesirable activities including masochistic operations.

Under no circumstances would my law permit the use of force, compulsion, or peremptory procedures to obtain body tissues from the living, no matter how worthy the purpose. It follows that many potential recipients will remain legally unaided. The result of the hypothetical case in Chapter 5 of the man who refused to give blood to the dying President of the United States would be the death of the President. In my opinion, the 1978 case of *McFall* v *Shimp*, in which a United States judge refused to order compulsory bone marrow removal from an unwilling adult, was rightly decided. These cases raise moral questions for which legal answers are not applicable. Society should attempt to solve the problems they pose by means other than legislation.

For living minors and mental incompetents, my law would also be consent-based, but more complex. I would build from the base that, as with normal adults, consent is necessary and that body-part removal by any form of duress will be unlawful. However, I would not accept the sophistries of consent which have characterized many of the United States judicial decisions discussed earlier in this book, and I would take steps to ban activities of a kind which are not unknown in other Western countries, where families will often overbear a child and hand him or her over to the surgeons to serve as a donor for a sick or dying relative. It is not acceptable in this field to apply the logic of substituted judgement or similar theories which have been used to justify removals. Their justification begins with the proposition that the minor or mental defective is legally incapacitated and therefore unable to give a consent to the surgery. It then runs along the line that the community, in the person of a court or an official guardian, is able to act in the consent-giving role. If it does so and gives permission, this will serve as the consent of the incapacitated person. I am unable to accept this argument, logical though it may be, because these decisions can be made regardless of whether the incapacitated person comprehends the issues. Often the subject, or victim, has no idea what the procedure is all about; certainly this is true of a four-year-old child or a grown mental deficient with a mental age of six.

The case for total prohibition of all forms of tissue removal from the bodies of those without legal capacity rests on the ground that they are incapable of consent by the law's own definition. I take the view that fundamentally, this is no more than an easy way out. Although not nearly so objectionable as the principle which allows courts to give consent on behalf of minors, it follows the same kind of unbending logic. It accepts without question the idea that every person under the age of majority and every person classified by law as mentally incompetent in all circumstances lacks the ability to reach an informed decision. Not only is this not true, it is not even good law, at least not good common law. It places the mature, intelligent seventeen-year-old on the same level of understanding as the idiot. Admittedly, a law of total prohibition would have the advantage of certainty. It would be known that no minor could ever donate any of his tissue to another person, and thus all minors and their families would be released from the agony of difficult decisions. But merely one hypothetical example will illustrate the tragic injustice which could be done by a blanket ban. Let us take Sue and Jenny from the example in Chapter 9 and assume them to be identical sixteen-year-old twins. Both are highly intelligent and live with their parents in an affectionate, close-knit family. Jenny's kidney disease will eventually be fatal unless a successful transplant is performed. The prospects of successful transplant of one of the healthy twin's kidneys are virtually 100 percent. No other kidney is currently available, and if it were, the best chances would be about 60 to 65 percent. Sue understands fully what removal of her kidney involves and has discussed it with her parents, independent doctors, her teachers, and her religious advisers. If she is prevented from giving her twin a kidney and the twin dies, the grief she and her family will experience will be permanent and incalculable. This kind of case is not as unusual as it may sound.

In my opinion, a law which flatly prohibited the removal of one of the healthy twin's kidneys for transplant to her sister would not only be unjust, but morally wrong. Inflexible rules should be avoided because they cannot be adapted to the infinite variations of human behaviour.

My law would first proclaim a prohibition of body-part removal from persons of diminished capacity, but it would not be completely rigid. The general rule would reflect the policy of the statutory provision. It would, however, contain machinery for allowing removal in certain cases. As far as regenerative tissues are concerned, it would permit removal (but only for the direct therapeutic treatment of another person) if the donor has the capacity to comprehend what is involved and agrees to it. In addition, parental consent and independent medical advice to parent and donor should be obtained. As for nonregenerative tissue, I would be much more restrictive. Not only should the donor be capable of understanding what is involved and agree to it, but parental consent plus reference to an independent ad hoc committee comprised of two or three members taken from the ranks of medical practitioners, psychologists, social workers, and judges would be necessary. Independent medical advice should be sought. In addition, for the same reasons given for adults, overriding restrictions would confine the purpose of the removal to transplantation within the donor's immediate family for life-or-death cases.

The purpose of this elaborate procedure is to allow no more than the most compelling deviation from the general rule. In my view, the principles could apply equally to both minors and mental incompetents.

My law would make no special protective provision for prisoners, institutional patients, and other persons in comparable positions of social disadvantage, such as members of the armed forces. Despite evidence of vulnerability to pressures to donate tissues or participate in experiments, there is not sufficient reason to classify such people in the same or a similar way as those lacking a degree of capacity. The requirement for a free consent will protect them and also apply to the behaviour of those with influence over them. There is no reason to treat such people as though they are incapable of making up their own minds about bodily integrity. A great deal of protection is already conferred by existing laws.

With the living, my law would reject nonconsensual tissue removal.

Compulsion is likely to be unacceptable to most people. With the dead, it is not so simple, because much depends upon the way one thinks of a corpse. In truth, we are not dealing with the bodily integrity of a human being at all, but the question for lawmaking remains the same: Must the community wait to be given, or may it take? The attitude of society to dead bodies will govern the kind of law which will be produced.

Although a widespread social belief may exist according to which people insist on retaining personal autonomy beyond the point of death until burial or cremation, I believe that this is not universal, and what is more, it is changing. The Australian model code of 1977 requires consent or nonobjection before the tissues of the dead can be removed. This requirement was a reflection of the lawmakers' interpretation of Australian public opinion. In principle, new laws should not simply mirror current public will, but for this subject, public opinion is unusually important. Many of the questions cannot be answered in black-and-white terms. My own opinion, reached after the Australian public inquiries of 1976 and 1977 and after long examination of Western attitudes as reflected in popular and academic writing and changing public laws, is that the public is by no means permanently committed to a single view of the dead. Increasingly one reads or hears in ordinary conversation on this subject the pleonastic but significant phrase "When you're dead, you're dead." What is meant by this statement is, "I don't mind what happens to my body after I'm dead."

A lawmaker must reflect on human feelings about the newly dead even if it is impossible to find satisfactory answers to all questions. It is arguable, for example, that the impulse to retain control over the bodies of people who have just died is often the result of earlier experience of grief and reaction to the deaths of family members. Possibly the most unacceptable aspect of the death of a close relative is the fact that he or she will never be seen again. Dead persons can be vividly pictured by the survivors, their comings and goings seen mentally, rather like the red lines on a long-exposure photograph of night traffic. Their voices and mannerisms can be recalled and their personal possessions touched. At the time of death, it may be difficult, even impossible, to believe that the person has from that moment vanished forever. This may explain the human desire to retain control of the

corpse for a short while and the outrage which is often shown at interference with dead bodies.

Having said all that, the truth is that dead bodies are soon destroyed, whether by natural decomposition or cremation. Organs and
tissues which can be used to save lives rot rapidly. It is therefore very
much to the point to ask whether an official insistence upon access to
dead bodies is really an attack upon personal autonomy. My feeling is
that in essence it is not. Certainly the dead must be respected, and the
public is right to insist upon this, but the dead body is a thing utterly
different from the living body. The very idea of applying the notion of
personal autonomy to a corpse is absurd; at most, personal autonomy
is only artificially extended beyond death. What we are dealing with
here are deep-rooted attitudes and nothing more, attitudes which are
in fact changing. This means that the law must be painstakingly
framed so as to preserve human ideas of dignity. A lot will depend on
timing. Is the community ready for this type of legislation?

My own law would move away from the requirement of consent
and embrace the idea of contracting out or community availability. I
believe that Western communities are prepared to allow the use of
their dead for the therapy of the living, provided that it is decently
done. The great value of human tissues is now established. I can see no
objection to a law under which the bodies of the dead may be
promptly used for public health purposes. Whether this position seems
to interfere with preserving personal freedom will depend upon
whether you believe freedom is something which should extend to
your corpse. I do not believe that it should, and moreover, take comfort from the fact that my law, along with others of its kind, will allow
everyone to signify an effective objection while alive. It is plain that
such a law should be accompanied by safeguards. These should not
only aim at forestalling public alarm but should protect citizens from
overenthusiastic medical practitioners.

My law would permit tissue removal from all dead bodies for the
purposes of therapy (which includes transplantation), or for other
medical or scientific purposes. My protective measures would be considerable. I would accept the European contracting-out principle
under which any person may record an effective objection during his
lifetime. Bearing in mind the depth of public feeling on this subject
and the inertia of most people, I would insist that widespread publicity

be given to the law and its objectives. There would be a policy of public education. People would be given easy facilities for recording objection.

The law should also compel the medical profession to behave with swiftness, decency, and humanity when exercising state rights over dead bodies. Penalties should be established which would placate public anger in cases of abuse of official power. As suggested in preceding chapters, the standing of the medical profession should influence the prescription of penalties. A community whose medical practitioners are of a high calibre and observe a recognized ethical code need not apply as strong an arm as communities that are less fortunate with their doctors. Even so, it may not be desirable to carry fine tuning to the lengths of the English Act and provide no sanctions at all.

The family of the deceased is ignored under a contracting-out law. This must be carefully considered because of the profound distress which is to be expected on the part of close relatives. Some families may be appalled at the thought of the dead member's body being subjected to removal of various parts. It is for this reason that I would place obligations on medical practitioners, and insist upon public education on the existence and purpose of the law. A complaint machinery could be devised if required, to provide some form of control over improper behaviour.

After legislating for body-part removal from the living and the dead, important but subsidiary matters remain to be dealt with. My law would give special attention to blood transfusion because blood is a tissue with special characteristics. In principle, it is no different from other regenerative tissues, but its painless, simple removal and unique lifesaving qualities would lead me to create clear rules for its acquisition and use. I would be prepared to simplify its removal from children (still with safeguards such as the child's consent plus parental consent), as well as its administration to children and incompetents even over the objections of parents and guardians.

Other subjects which merit legislative clarification are the need for privacy between unrelated donors and recipients, and confidentiality of medical records. The question of legal liability of doctors and medical staff is also significant. Those who act honestly and in good faith should have protection from unjustified complaints.

I would prohibit commerce in human bodies and body parts but would allow the payment or reimbursement of expenses incurred by a donor. The prohibition of commerce would be worded so that the processing of tissues by laboratory methods and the charging of proper costs for that service would not be prevented. As a salutary reminder, I would bear in mind that commercial traffic can be a two-way affair. Occasional reports appear which show that tissue recipients are quite capable of looking a gift horse in the mouth. In 1971 in New South Wales, Australia, a young rodeo rider was fatally injured when thrown from a horse. When death was imminent, his parents agreed to give his kidneys for transplant. The young man was then transported by ambulance to a city hospital where he died and his kidneys were removed. The body was later returned to the parents with a bill for transport and medical fees amounting to $A 550, or approximately $U.S. 630.

My law would specifically recognize the medical determination of human death by reference to cessation of brain function. The diagnosis of brain death would be legally accepted so that there would be no possibility of accusing doctors of homicide, as occurs in England with regularity. There are few more unpleasant or bizarre sights than the recurrent English newspaper headlines which ask whether the patient who was stabbed, raped, kicked to death, or battered with a brick was killed by the assailant or by the intensive care specialist who pulled the plug. The criticism that headlines of this kind call for should be directed, not at the newspapers, but at those who are in a position to bring about law reform and fail to do so. Because there has been no authoritative legal pronouncement by statute or the appellate courts, coroners' inquests and criminal trials have time after time raised the question of the doctors' liability. Fortunately the anwer has always been in the negative. Perhaps also fortunately, for politicians, no measurement can be made of the influence of these events upon the public's willingness to donate tissue, the medical profession's support for transplantation, or the death rate of patients in need of organ transplants. Legal uncertainty stultifies transplant programs, causes unnecessary deaths, and frightens doctors and hospital administrators.

For reasons appearing in the preceding chapter, my law would contain no direct regulation of the recipients of body parts. The legislation would be as brief and clear as possible, with sanctions and penalties designed to ensure its efficient operation.

I would try at every stage to maintain a balance between the individual and the group, and to recall that my work will make a new disposition of the human body. Thought on long-term implications as well as immediate effects is necessary, in particular on the possible conjunction of the new law with other rules that may be required for emerging practices in human reproduction and genetic engineering. Whatever our metaphysical attributes may be, the only thing that identifies each one of us as a human being is the body.

Index

The human body is not what it used to be. Spectacular new advances in medicine have initiated an era in health care that a short time ago would have qualified as science fiction. People who once would not have survived illness or injury are leading normal lives today thanks to other people's blood, organs, and tissues, as well as plastic and metal valves and limbs. Babies are begun in test tubes. Death has been redefined, and society has a natural source of donor parts ("brain-dead" corpses whose organs are still functioning).

But these boons for medical science may not augur well for personal autonomy. Once the human body has value as a stockpile of curative materials, the problem of ownership arises. The perpetual confrontation between society and the individual takes on a new aspect: it is no accident that the only precedent for humans being treated purely as property is slavery.

Russell Scott explores both the dark and the hopeful sides of this major social issue. He traces the history of medicine's demand for corpses since the days of the early anatomy schools and cites case after modern case, heard in courts throughout the world, in which the inflammatory matter of rights to other people's bodies has been raised. The results of these lawsuits do not give us a consensus, and doctors, patients, and jurists disagree about what to do next. It is clear that human organs are needed for research and transplantation, but from whom are they going to come? And how are they to be obtained? By consent? Commerce? Compulsion? Some governments are already asserting territorial claims on their citizens. In short, do you own your body, or does the community? In answering this question, Russell Scott argues persuasively and eloquently that progress in treatment of the sick cannot be bartered for human liberty.